Human Trials

Human Trials

SCIENTISTS, INVESTORS, AND PATIENTS IN THE QUEST FOR A CURE

Susan Quinn

A Merloyd Lawrence Book
PERSEUS PUBLISHING
Cambridge, Massachusetts

Cataloging-in-Publication Data is available from the Library of Congress.

ISBN 0–7382-0182-0

Perseus Publishing is a member of the Perseus Books Group.
Find us on the World Wide Web at http://www.perseuspublishing.com

Perseus Publishing books are available at special discounts for bulk purchases in the U.S. by corporations, institutions, and other organizations. For more information, please contact the Special Markets Department at the Perseus Books Group, 11 Cambridge Center, Cambridge, MA 02142, or call (617)252-5298.

Text design by Tonya Hahn
Set in 12-point Fairfield Light by Perseus Publishing Services

First printing, April 2001

1 2 3 4 5 6 7 8 9 10—03 02 01

Contents

To Esther Taft Quinn, who has always been my first reader and best example

Acknowledgments

Throughout this undertaking, I have relied on the generosity and trust of researchers, clinicians and patients, as well as the employees of AutoImmune, Inc. Many are named in the book. Among those who are not, I am especially thankful to Patsi Nelson, the scientist who first welcomed me to AutoImmune, and to Byron Waksman, a distinguished researcher and clinician who provided me with a rich introduction to the history of autoimmunity. Neurologist Norman Kachuck and research coordinator Kathy McCarthy, at the University of Southern California MS Comprehensive Care Center, provided important information about clinical trials, as did Helen Adamson, the clinical research coordinator at the Gunderson Lutheran Medical Foundation in LaCrosse, Wisconsin, and Rozie Arnaoutelis, who plays a similar role at the Montreal Neurological Institute. Finally, I am grateful to all the patients I spent time with for their willingness to allow me into their lives, in good times and bad.

Thanks also to Bernadette Bensaude-Vincent, Les Boden, Marnie Mueller, Constance Perin, Pat Thomas, Jan Schreiber, and Deborah Weisgall, all of whom read this book at various stages of its development and offered important advice. Thanks also to Evy Davis, Kathryn Kirshner and Diana Korzenik, who have read some parts more than once, and frequently helped strengthen my resolve, and to Joyce Antler, Fran Malino, Megan Marshall, Judith Tick and Lois Rudnick, who have lent their astute ears. I feel fortunate, too, to have worked with Merloyd Lawrence, a fine and seasoned editor, and to have had the guidance of Georges Borchardt. The reader whose opinion matters most to me, and whose love sustains and inspires me, is my husband, Dan Jacobs.

Note to the Reader

THE QUOTED WORDS IN THIS BOOK come from my interviews and observations of events, from tapes and videos of events I could not attend, and from the journals of Howard Weiner, whose record keeping has provided a window into a world that is usually inaccessible. To protect the privacy of some patients, I have changed their names. When the first name or first name and last initial are given, the name has been changed. When both first and last names are given, the name has not been changed.

Human Trials

CHAPTER 1

The Clinic

MOST OF NEUROLOGIST HOWARD WEINER'S DAYS are spent in his laboratory or on the road, talking at conferences. But on Tuesdays, he takes the white coat, size 42 long, down from his office door, pulls it on over his blazer, and walks from his lab to the multiple sclerosis clinic at Brigham and Women's, a Harvard teaching hospital in Boston.

On this particular Tuesday, it's windy and cold as he heads across the mall of Harvard Medical School, his white coat flapping, at a walk that verges on a trot. Inside the hospital, he swings by the coffee shop and picks up juice and a blueberry muffin to be consumed between consultations, and hurries down the long corridor everyone calls "the Pike."

On the Pike, which connects the old and new Brigham hospital buildings that have accumulated over a century, Weiner slows momentarily to meet and greet other white coats moving in the opposite direction. Then he hurries across the grand hospital lobby and through the neurology waiting area to the doctors' room, where he scans the roster of multiple sclerosis patients he will see that morning. As usual, he recognizes most of the names on the list: Many of them are long-term patients, and most of them are not doing very well. "There are some very sick people coming in today," he says softly.

Howard Weiner is a fit six-footer with thick glasses that can make his eyes look wild. His hair used to be abundant and curly, but now that he's in his fifties, it's slicked down and thinning on top, adding prominence to his nose. He has a gleaming smile that flashes often, but he is incapable of irony or sarcasm. Because of his deep bass voice and his earnestness, one of his poker buddies thinks he would have made a good rabbi. Weiner, however, has chosen to devote himself to another mission: finding a cure for multiple sclerosis. Even after thirty years of searching, including many setbacks and frustrations, his zeal is undiminished.

Weiner has always been an activist when it comes to treating multiple sclerosis. "I think you've got to be doing something," he frequently declares. To buttress his argument, he sometimes quotes a line from *King Lear*: "Nothing will come of nothing." But the truth is that, much of the time, multiple sclerosis defeats his interventions. That's why, on Tuesday evenings, he so often writes in his journal about how "tough" his day in the clinic has been. He doesn't like seeing people getting worse.

The first patient on the roster this Tuesday morning is new to him. She is Harriet M., a forty-three-year-old from New Hampshire who is experiencing numbness in one arm and has come to get a second opinion. Harriet is a small woman with curly, bleached hair that creates a halo around her face. She is wearing comfortable knit pants, which accommodate her slender waist and full, low hips, and a short T-top reveals a little of her firm midriff. Her eyes look puffy and tired, but her skin is smooth and youthful, and her arms look strong. She lifts weights, she tells Weiner, takes care of herself.

There are two children—a son out in Las Vegas, "sowing his oats" working as a blackjack dealer, and a fifteen-year-old daughter still at home. And there is a husband, sitting there next to her with his arms folded across his round belly, outwardly calm. The neurologist in the small town where they live wasn't entirely sure whether she had MS or not, she explains to Weiner, and wanted to wait a couple of months before starting any treatment. Their son, worrying on the phone from Las Vegas, said he thought that the doctor was too nonchalant. So the two of them took the day off from work—she has a clerical job, he runs a news distribution service—and started out for Boston before dawn on this brisk November morning for an appointment with Howard Weiner, an expert who can clear things up, for better or for worse.

Weiner listens to Harriet's story, a variation on hundreds of others he has heard in the MS clinic over the past twenty-five years. Eight years ago, she had a cold that resulted in dizziness and numbness on one side of her face. Her doctor referred her to an ear specialist, guessing that she probably had an inner ear disturbance related to the infection. But for two years after that, when she looked at her feet, she got a strange sensation in her back. When her left arm began to get numb recently, she thought it might be because she had carpal tunnel syndrome from her job. That was when her doctor mentioned MS as a possibility.

The story has revealed a lot to Weiner. Probably, the facial numbness eight years ago was a first MS attack. Most telling is the report that Harriet has been feeling shocks down her spine for the last two years when looking at her feet. This was probably the "Lhermitte's sign," caused by a lesion in the spinal cord that increases pressure when the neck is bent forward and down. It is a classic symptom of multiple sclerosis. So far, though, Weiner is going only on the patient's history. Now he asks her to sit on the end of the examining table so that he can run through the neurologic exam.

Howard Weiner likes to come prepared: His family teases him that he never comes home to Colorado without a set of clothes for every kind of weather. He always carries an extra laser pointer with him when he's giving a talk, just in case. Similarly, Weiner keeps his white coat fully equipped for clinic days. On one breast pocket is the white plastic tag bearing the wavy crest of "the Brigham," next to the words "Howard Weiner, M.D., Neurology" and inside there's a plastic eye chart and a tuning fork for testing sensations. A black rubber reflex hammer and stethoscope weigh down the lower right pocket and there's a prescription pad, lower left. His car keys, also essential, are dropped in at the last minute.

Now he taps Harriet's knee with his reflex hammer. He runs his car keys along the arches of her feet, noticing that her right big toe turns up instead of down, as it would in a normal subject. He asks her to follow his finger with her eyes, looking for "disconjugate eye movements," which indicate that the brain stem has been affected. Finally, Weiner watches Harriet walk, a slight bit unevenly, from one end of the corridor at the Brigham clinic to the other.

By the time he's finished the neurologic exam, he strongly suspects that Harriet has an MS lesion affecting the left side of her body. But he won't say anything until he's taken a look at the films she brought along. He carries the giant manila envelope down to the doctors' room and throws the stiff films up, two at a time, onto the view box. Very quickly he finds a white spot on the vertical ropes of gray that tells him she has a lesion on the left side of her spinal cord, in the neck area, then two other lesions in the brain. To Weiner, the conclusion is now unavoidable: This woman has multiple sclerosis.

Diagnosing MS is tricky, because the symptoms are various, and they come and go. But for a long time, diagnosis was about the only thing neurologists could do. "For a neurologist," writer and MS patient Nancy Mairs wrote in 1989, "MS must be the worst possible fate, worse even than a brain tumor," which offers a "chance for heroic rescue. . . . With MS, they stare powerlessly, sometimes for decades, at inexorable degeneration."

Until the 1990s, there were no drugs specifically designed to treat MS, and very few that helped. Steroids could be given to calm inflammation, but they had no effect on the progress of the disease. "Neurologists described patients and diseases and wrote atlases about them," says neurologist Larry Levitt. "But they didn't *do* much. They were very nihilistic about multiple sclerosis—they'd say come back in six months and I'll recheck to see if you have even further trouble walking."

Levitt, who practices now in Allentown, Pennsylvania, was Howard Weiner's chief resident when the two of them trained at Harvard. He has watched his old friend Weiner challenge this nihilistic attitude about MS in his research and treatment, sometimes incurring harsh criticism from colleagues. When Weiner initiated an MS trial of a drug previously used in cancer treatment, back in the late seventies, many in the field faulted him because the underlying mechanisms that caused the disease were unknown. "He recognized and admitted the fact that it wasn't the best way to do it," Levitt says, "but it offered hope and was effective in some cases."

In recent years, Weiner has had a lot of company, as neurology has moved from passive observation to active intervention. MS trials abound, even though the disease mechanism is still only partially understood. When the American Academy of Neurology gathered in Boston in 1997, the meeting had a title that reflected this change. "Revolution in Neurology" appeared in bold letters on the thick program, with the R doing double duty in the symbol Rx. By 2000, there were about 100 clinical trials planned, in progress, or recently completed.

Some successes have emerged from all this activity.

There are three drugs on the market that target MS: Avonex, Betaseron, and Copaxone. The ABC drugs, as they are called in MS circles, have given new hope to patients, but they are not the cure. The drugs can reduce the MS attack rate by only about 30 percent, and they don't work in all cases. All three must be given by injection on a frequent basis, and there are sometimes side effects. In addition to the ABC drugs, chemotherapy agents are gaining acceptance for the treatment of certain forms of rapidly advancing MS. But these, although sometimes dramatically effective, have worse side effects than the ABC drugs.

About fifteen years ago, Howard Weiner got an idea for an entirely new and potentially revolutionary approach to multiple sclerosis. This approach, if successful, could result in a safe, oral treatment not only for MS but also for other diseases with similar autoimmune characteristics. From the moment he discovered that the approach worked in animals, he believed its development was going to be his crowning scientific achievement. The idea, called "oral tolerance," was ridiculed in the beginning, but it has since become the focus of research in laboratories all over the world. So far no one has been able to show conclusively that it works in treating human disease. "Sooner or later," Weiner insists, "it has to work. When that happens, I'll be a happy man."

For the New Hampshire couple who have come to see him on this Tuesday, however, such dreams of future breakthroughs are no comfort. Weiner enters the room where Harriet and her husband are waiting and hoists himself onto a swivel stool. With his hands clasped around one knee, he turns to face the couple. When you have only one symptom, he begins in his low, ponderous voice, then it's difficult to know if it's MS. But when a number of symptoms point in the same direction, as in this case, it becomes possible to make a diagnosis. Then, incongruously, Weiner offers the New Hampshire couple a big-city analogy: When you see the Empire State Building *and* Central Park *and* Madison Square Garden, he tells them, you know you're in New York.

Neither of them smiles, or even seems to hear what Weiner has just said. But then, as he starts to speak again, the woman stops him. "Wait a minute," she says in a trembling voice. "You say you *think* I have MS. It's not for sure?"

"No," Weiner answers firmly. "It is MS. I have no doubt."

He continues. "MS is an autoimmune disease of the nervous system in which some of your cells attack your own tissue. You probably had a first attack eight years ago, and that's lucky, because there has been a long interval between attacks. But with MS, it's impossible to know the future. It's a little like being in a boxing ring where you keep getting hit by an invisible opponent. You may be in the ring with a five-year-old who doesn't hit too hard. Or you may be in there with a pro who does."

None of this is registering with the patient, and only a little with her husband. "Do you understand?" Weiner asks. "Well, right now I . . ." her voice trails off. She is staring at the floor, about to cry. Understanding will be a long and complicated process.

Back in the doctors' room, Weiner picks distractedly at the dome of his blueberry muffin and washes it down with juice as he dictates his orders into a small recorder. He recommends that Harriet start immediately on a course of injections of one of the ABC drugs, combined with a course of intravenous steroids to calm the current inflammation. Before he moves on, he tells MS clinic nurse Lynn Stazzone that the patient is "scared" and needs to talk more about what she is facing.

In his research Howard Weiner accepts, reluctantly, the slow pace of progress. But when a meeting goes on too long he can be seen glancing at his watch. When it comes to sitting with patients in distress, he relies on Lynn Stazzone. Lynn is famous for taking as much time as necessary. One of her patients confides that she sometimes tires of waiting for Lynn, who is often late, but then her turn comes, and she knows Lynn is completely available to her for as long as she needs her. Then she forgives her.

Before she leaves on this day, Harriet will talk over the meaning of her diagnosis with Lynn Stazzone and she will be loaded down with literature. After she leaves, if she is like most MS patients, she will continue to investigate with a fury. She will do what one patient called "going to Dr. Library." She will contact the MS Society, which provides information about support groups and MS walks. She may well log on to the chat rooms on the Internet, where hours are spent comparing symptoms, treatments, and strategies. In the end, she will be forced to conclude that MS is a frightening and capricious disease.

MS is rarely fatal. But future attacks may result in vision problems, speech impairments, and incontinence, and occasional diminished intellectual capacity, as well as disabling paralysis that can affect arms as well as legs. Walking is likely to become difficult as the disease progresses, and she could eventually end up in a wheelchair, unable to use her legs at all.

Of all the afflictions of multiple sclerosis, the most difficult may well be the disease's unpredictability. In Harriet's case, as Weiner noted, it is a good sign that she had her first attack eight years before. On the other hand, the eight-year interval is no guarantee of anything. Many patients whose disease has been mild for years take a sudden turn for the worse, and end up with paralysis and other distressing symptoms. In medical parlance, a disease that is "relapsing and remitting" can become, for no apparent reason, "progressive." So when the patient asks that most important question, "What's going to happen to me?" the doctor can only answer, "I don't know."

Sometimes there are happy surprises, as inexplicable as the unhappy ones. Waiting for Weiner in another examining room is Natalie H., a sixty-five-year-old with the mildest possible MS. Coming to see Weiner, as she has for many years, means taking a break from a life of travel. Natalie is trim, tanned, well coiffed. Her hands are manicured and adorned with gold bracelets. As Weiner goes through the motions of the neurological exam, she talks about her golf game: She still walks well enough to keep it up, she tells him, though she has some numbness in her hands, which affects her putting. She treats Weiner as the young doctor he was when their relationship began. She teases and he teases back. Weiner is grinning as he leaves her.

The rest of the patients Weiner sees this day are not so lucky. Carol S., a woman of about sixty, transfers her weight from a wheelchair to a walker and makes her way, with great effort, along the corridor. Given a choice, Weiner himself always walks very fast. So it is ironic that he is the inventor of something called the "ambulation index," which measures the often slow pace of MS patients. The index allows him to assess changes in walking ability, from one appointment to the next. On clinic day his father's Rolex, which usually moves him briskly through his daily printout of appointments, is put to a different use—timing patients as they walk, slowly and often with great difficulty, the length of the clinic corridor. The ceiling panels above serve as his unit of measurement. Carol S. seems to think she's moving better than last time, but Weiner, glancing from his watch to the ceiling panels, has a slightly pained, doubtful look on his face. Both doctor and patient know there isn't very much that conventional medicine can do for her anymore.

But Carol has brought a sheaf of papers describing a nutritional program that she hopes will help. It has to do with someone's complicated theory about the connection between MS and allergens and the blood-brain barrier, and it requires a number of unusual lab tests. Weiner agrees to look into it. In fact, this will be yet another task for the MS clinic nurse, Lynn Stazzone. Minutes later, she is on the phone with the lab people, trying to explain her reasons for needing tests the lab has never heard of.

"I think you have to do everything you can to support people," Weiner says. "Many of them are trying to put some order onto their life, onto their disease, trying to find some way to help themselves. I never ridicule them for crazy things." But the fact that so many patients are seeking alternatives is yet more evidence that Western medicine has weak answers when it comes to multiple sclerosis.

Women are twice as likely as men to get MS. So it is not surprising that there is only one man on Weiner's roster this morning. He is a doctor, an anesthesiologist whose disability has made it impossible for him to work at his specialty. He attempts the ambulation test with a walker, but it's so belabored that Weiner stops him. He is only fifty, but he looks older, pale, and unwell, sitting in his wheelchair. He agrees to a new treatment, hoping that it will work better than the ones he's tried so far. "I think he was going down like this," Weiner says, bending his hand toward the floor, "and we maybe changed the angle of his decline a little bit. Dramatic impact we haven't had."

The last patient of the morning is the most desperate. There isn't any chance this woman, who is in her late forties, could take the ambulation test. Her legs are pencil-thin from disuse, her head is fixed to one side, and there are hesitations in her speech. She is wearing nice clothes: a matching sea-green skirt and blouse and a little black jacket. But everything in her life seems to be going from bad to worse. She's battling incontinence (an MS symptom) and she smells of urine. She has had an infection that put her in the hospital with something close to toxic shock. She took a leave from her job two months earlier without pay. Yet she still hopes to go back to work, and she asks Weiner over and over about new drugs she could possibly take. Most of them aren't possible, as Weiner explains with studied patience. Weiner agrees to try a low dose of methotrexate, an immune suppressant, but only after a liver test determines she can handle the side effects. "I'd like to try something," she tells him, the tears welling up, "because I'm definitely getting worse."

Outside the examining room, Weiner admits that the methotrexate probably won't help this woman. Once someone has been in a wheelchair for a while, it's too late to do much. The earlier you treat the disease process, the more likely you are to have an effect. But even among those in the early stages of MS, there are responders and nonresponders.

The same unpredictability that makes MS so hard to treat makes it difficult to study. There is simply no usual course for the disease: Some cases are mild, some severe, some progress rapidly, and in some there are long lapses between attacks. How is it possible to measure the effectiveness of a treatment, if every MS patient's disease behaves differently? The difficulty of this task is one of the reasons a highly effective treatment for MS has been so elusive. But then, as Howard Weiner points out, none of the diseases that still confound biological science are easy or uncomplicated. "The low-hanging fruit has been picked."

CHAPTER 2

The Disease

THE FIRST PERSON TO DESCRIBE CLEARLY the symptoms of multiple sclerosis was a member of the English aristocracy, Augustus d'Esté.

Augustus was the grandson of King George III. But because the king didn't sanction the match from which he issued, he was never, despite a lifelong effort, granted royal recognition and rank. This was one of the two great difficulties of his existence. The other was what he referred to, in the journals he kept throughout much of his life, as "my infirmity."

The evidence suggests that Augustus was not a particularly admirable character: "extravagant, careless, and selfish," in the words of Douglas Firth, author of a brief biography. As a student, he borrowed money from the servants, and formed a liaison with a housemaid. "The gratification of his every wish and whim by an indulgent and adoring mother can only have enhanced these faults," notes Firth. Yet the self-centeredness of Augustus d'Esté has benefited posterity. It is because he was so fascinated with his person and with the changes in his body, and because he wrote of them so precisely, that we have a vivid picture of a case of multiple sclerosis from the early nineteenth century.

Like many MS sufferers, Augustus was in his twenties when he noticed his first symptom. He had traveled to the Scottish Highlands for the funeral of a dear relative. When his vision blurred, he attributed it to weeping, especially since he noticed soon after, when arriving in Ireland, that his eyes were restored to normal. In fact, he had experienced a fairly common first symptom, double vision, resulting from a lesion in the brain stem.

Three years later, while he was in Florence, Augustus's eye problems returned. "The malady increased to the extent of my seeing all objects double," he wrote. Then, no sooner had the double vision abated than a "new disease," as he called it, began to appear. "Every day I *gradually* felt my strength leaving me." It became more and more difficult for Augustus to go up and down stairs, even though it helped momentarily if he "slapped myself sharply on the loins." One day, he fell twice when attempting to reach the bathroom. "I was obliged to remain on the floor until my Servant came in and picked me up." He continued in this weak state for three more weeks, falling down on several occasions "from my legs not being strong enough to carry my body." At the time Augustus d'Esté was thirty-three years old.

His disease was following a typical pattern: an attack, followed by a remission, then a more serious attack, followed by another remission. But also typically, as time went on, the remissions did not return Augustus to his original state of health. He improved dramatically after the Florence attack—an improvement he ascribed to his new doctor's orders to eat two beefsteaks a day and drink plenty of London porter and sherry. Traveling on to Rome, he was able to walk the steep hills of the city and ride horseback. But he was "never able to run so fast as formerly, nor could I venture to dance." There were other symptoms: problems in urinating and spasms of pain in his feet and legs. He reported sexual difficulties as well, what he called "a deficiency of wholesome vigor in my acts of connection."

Augustus traveled the Continent in search of remedies and consulted many doctors. Some of their recommendations were harmless if not helpful. He immersed himself in baths of various temperatures and in the frigid ocean, and took long walks and very long horseback rides. He also subjected himself to bleeding with leeches and ingested prescribed potions with sometimes alarming contents, including strychnine and mercury. Every once in a while, he would dare to protest. "In humility," he wrote at one point, "I attribute this attack to . . . the five preceding prescriptions on the brain." After a treatment of "electrification," he wrote that "it is clear and apparent to me, that Electricity is the most powerful Agent to my injury instead of to my recovery." For the most part, though, Augustus followed his doctors' orders with touching fidelity.

Yet despite his best efforts, the disease continued its gradual debilitating course. By his forties, he was unable to walk "without a Stick." At age forty-nine, while riding his phaeton between Windsor and London, he once again lost all strength in his limbs and had to be lifted out of the phaeton and carried to bed. Two years later, his left ankle became so weak that he had to use a "steel upright." By his fifties he was dependent on a "chair on wheels." Until he died at fifty-four, he made a great effort to walk when he could, and kept a record of the number of times he was able to circle his room. But, as he admitted, "I lie sadly much on the Couch."

Twelve years after the death of Augustus d'Esté, a drawing appeared in several medical atlases depicting plaques, dense areas of scarring, in the central nervous system. Eight years after that, the brilliant French clinician Jean-Martin Charcot connected such scarring to a catalog of symptoms that included paralysis of the lower extremities and double vision, differentiating it from Parkinson's disease, which had been described forty-four years earlier. Because of the scarring seen in the brain, Charcot called the disease "sclerose (scarring) en plaques," which was translated into English as "multiple sclerosis."

After Charcot, other neurologists sharpened the definition of MS, and described its effects in more detail. Scarring, it turned out, was a secondary event: The main event was the inflammation that preceded it. There was also, as Charcot had suggested, a peculiar quality to the inflammation: It occurred in the sheath surrounding the spinal cord rather than in the cord itself. The sheath is made of a mixture of lipids and proteins called myelin, which wraps around the spinal cord in much the way insulation covers electrical wire. When spots in this coating are eaten away, the bare places cause a disruption in the flow of signals to the brain.

The way MS affected the myelin sheath was understood by the middle of this century, but the underlying cause was a mystery. It might have remained so, had MS continued to be seen narrowly as a neurological disorder. But in the 1940s and 1950s, a revolution took place that cast multiple sclerosis in a whole new light. Nowadays multiple sclerosis is seen not just as a disease of the nervous system that affects 350,000 people in the United States, but as one of a family of autoimmune diseases that affect millions. The revolution that caused this new understanding took place not in neurology but in immunology. At the center of it was the work of two Rockefeller Institute scientists, Karl Landsteiner and Merrill Chase.

The office Merrill Chase occupies at the Rockefeller Institute is so narrow it might once have been a storage closet. There is barely enough room for a desk, a chair, and a bookcase on which to stack his many articles. Chase, now ninety and bent to nearly a right angle at the waist, rises from his chair to pull an article off the pile with his long, pale fingers. On this day he is dressed impeccably in a gray pin-striped suit and maroon tie with white polka dots. His carefully parted hair shows the marks of the comb he keeps in a crumbling leather case in his breastpocket. Merrill Chase will tell you that he has to struggle just to get from one day to the next: to maintain his ailing wife, who needs round-the-clock care, to pay his bills, and to keep coming to work. "What you see when you look at me," he says, "is glue."

Merrill Chase is a hero to many in his field, a researcher of great integrity and an immunological pioneer. Given his accomplishments, there is no need for him to keep coming in to the Rockefeller Institute anymore. But perhaps he does so in part to emulate *his* hero and mentor, the Austrian-born Karl Landsteiner, who was stricken with a heart attack while at work in the lab and died two days later. At the time of Landsteiner's death in 1943, Chase observed that one "should not deplore a quick merciful death coming in the midst of full activity after the lapse of seventy-five years of life."

In the 1890s Karl Landsteiner was starting out in Vienna and Louis Pasteur was ending a lifetime of research in Paris. Both were working in the field that would come to be called immunology, but their researches were aimed in different directions. Pasteur had discovered, in working to rescue the French silkworm industry from a devastating blight, that disease is the result of infection from a microorganism. His "germ theory" led him to the conclusion that vaccination, pioneered by Edward Jenner in treating smallpox, could be put to broader use. It was possible, for instance, to prevent cholera and rabies by introducing a weakened version into the body before disease threatened. The reason, as Pasteur and others were able to demonstrate, was that the blood, when prompted, produced antibodies to these external enemies. Even more important, the blood, or more precisely the B cells in the blood, had the remarkable ability to remember previous attackers and strike them even harder when they tried to invade again.

At the same time Pasteur was focussing on microorganisms outside the human body that could cause illness, Karl Landsteiner was studying differences within. What made blood from one human incompatible with blood from some others? In a series of studies started in 1901, Landsteiner showed that humans could be divided into several distinct blood groups. His work, for which he won a Nobel Prize in 1930, served as the basis for establishing the ABO system of blood types, and ushered in the modern era of blood typing and transfusion. Because of its great practical benefit, Landsteiner's work on blood types overshadowed his later accomplishments. But it was only the first step in a lifelong quest for answers to the question of how the body distinguishes self from nonself.

At the Rockefeller Institute, where he worked with Merrill Chase, Landsteiner focused his attention on allergic reactions, or hypersensitivity. For some time, scientists had been puzzled about a skin reaction in former TB patients that didn't fit Pasteur's antibody paradigm. When such patients were injected with a protein extract of TB, they would mount a reaction. The skin in the area of the injection would become red and itchy. This was thought to be an allergic reaction of some sort, caused by antibodies formed in the blood. But when an attempt was made to transfer the allergic reaction to a second person by transferring blood, nothing happened. Chase and Landsteiner, in a milestone experiment conducted in the 1940s, demonstrated that such hypersensitivity could be transferred to a second animal from a first, but not by transferring whole blood. Rather, they transferred cells (lymphocytes) extracted from the abdominal wall of the hypersensitive animal to a second animal and got a reaction. This was one of the experiments that marked the beginning of a major new subdivision of immunology, called cellular immunology. In time, cell-mediated responses would come to be recognized as the basis for many important bodily defenses, including the suppression of viral diseases and control of tumors. The cells that Landsteiner and Chase transferred have come to be known as T cells because of their origin in the thymus, a gland once thought to be useless and now considered by many to be the master organ of the immune system.

T cells are now understood to be a key component of immunity, at least as important as antibodies, or B cells, to human health.

The complexity of cell-mediated immunity has turned immunology into one of the most exciting and dynamic fields of biological research. The action of B cells is quick, specific, and relatively uncomplicated. The action of T cells is vastly more complex, and involves the activation of a cascade of signals, all of which have acquired names. At least 300 separate biological activities, by one count, have been ascribed to activated T cells. This has resulted in what one researcher called "an immunological tower of Babel." Some biologists have been more creative than others in meeting the nomenclature challenge. One researcher, upon discovering a tumor cell that behaved contrary to his expectations, labeled it the "FM cell," short for "fuck me." Other enzymes and factors in cellular immunology have been assigned numbers in arbitrary fashion. "Someday," as one researcher pointed out, "there will have to be a conference in some beautiful place, with splendid food, where all these names get sorted out."

But the naming of the cell-mediated activities is minor compared to the challenges that they have presented, over the last five decades, to old assumptions about how the immune system operates. The immune system, in the past, had been assumed to be the body's ally, and boosting immune response (as with vaccination) was understood to be a good thing. Furthermore, scientists had accepted the doctrine laid down by Paul Ehrlich, the great German biochemist, of *horror autotoxis,* a horror of self-poisoning. The body, in other words, could not be made to attack itself.

But this is most decidedly not the case. The body can and does attack itself at times. The immune system is not always an ally. In the large number of individuals who contract autoimmune diseases, the T cells mistakenly deem the body's own tissue to be foreign and mount destructive reactions. In fact, as immunologist William R. Clark has put it, "Our immune systems are like a highwire balancing act. Science and medicine have given us the means to keep our balance for most of the length of the wire, but it is still a very risky act." Many diseases, including not only old scourges like tuberculosis, but also new ones like AIDS, result when the immune system is inadequate. But when the immune system is too active, in an inappropriate fashion, other, equally devastating diseases can result. MS is just one of the autoimmune diseases, including diabetes and rheumatoid arthritis, that affect 8.5 million people in the United States alone. Increasingly, researchers believe that understanding and subverting the abnormal T cell action in one of these diseases is going to lead the way to treatment and cure for the rest.

CHAPTER 3

The Lab Director

HOWARD WEINER'S OFFICE ANCHORS ONE END of the lab at the Center for Neurological Diseases. It is the largest office, and one of the few with two doors. When he is out of town, the door to the lab is left open, so that researchers can come in to borrow books and journals, leaving a record of what they've taken on the white board. The bookcase occupies an entire wall of the office. On the top shelf are the black looseleaf notebooks Weiner has kept over the years, a record of his lab's search for a treatment for MS and other autoimmune diseases.

On the door hang the coats of the various roles Weiner plays: the long white coat for Tuesday morning clinic, the standard-issue blue blazer and striped tie for meeting the public, and a big green down jacket, in case he gets caught by bad weather. There is plenty of room, during his frequent absences, for others to spread out at the large conference table, but no one would think of doing it, or of settling into his chair behind the big oak desk. He is, after all, the boss.

Opposite the door, over the conference table, are a cluster of aphorisms, printed in bold letters and framed in black. One of them is from the speech Francis Crick gave in 1962, when he accepted the Nobel Prize in Biology for his discovery, with James D.

Watson, of the structure of DNA. "Politeness," Crick said, "is the poison of all good collaboration in science. The soul of collaboration is perfect candor—rudeness if need be. Its prerequisite is parity of standing in science, for if one figure is too much senior to the other, the serpent politeness creeps in."

The Crick quote is a particularly shrewd choice for the wall of a lab director's office. It states the ideal: "perfect candor." But it also suggests the problem: the serpent politeness, which creeps in if there is not "parity of standing." The Weiner lab, like every lab of more than a few people, is a hierarchy. The result is that Weiner must battle not only the serpent politeness, but a whole nest of other vipers—envy, resentment, territoriality. Competition, sometimes healthy, can turn into hostility that poisons the atmosphere. And perfect candor, when it comes from the director, can devastate a vulnerable young postdoctoral student. "Researchers," Weiner will tell you, "are incredibly fragile, incredibly vulnerable, because they're really out there."

Howard Weiner knows all this because he was once, like every lab director, a postdoctoral student himself, and he uses the good and the bad memories of that experience frequently, as he tries to create the ideal atmosphere in the complex institution he now leads. The Center for Neurological Diseases employs about 140 people. Half work on Alzheimer's disease, under the leadership of Weiner's codirector, Dennis Selkoe, and the other half work on autoimmune diseases, led by Weiner. Yet, even though Weiner gives the lab general direction and generates funds to keep it going, there are six semi-autonomous working groups studying autoimmune disease, each led by a principal investigator, or "PI," who obtains his or her own grants and oversees the resulting research. At the same time, Weiner is a PI himself, developing ideas for his own working group. As director and as a PI, he struggles to keep the lab's work coherent. In his journal, he often reminds himself that the key to success is "focus." The focus of the Weiner lab these days has widened to autoimmune diseases in general. But that is only because that is where the answers lie to Weiner's search for the causes and cure for MS.

Howard Weiner was intrigued by MS when he first encountered it in medical school. Then, in 1971, when he was working as a neurology resident at the Brigham, he was put in charge of a young man who had been hospitalized because he was having an MS attack. "It was very clear he was in trouble," Weiner remembers. "He wasn't walking well." Like most residents, Weiner turned to the literature to find out what to do, and discovered there was virtually no recourse beyond steroids to calm the inflammation. It was at that point that he decided to devote his career to finding a cure for multiple sclerosis.

Perhaps what drew him to this particular mystery was the young man's situation, which so closely mirrored his own. The patient was in his early thirties and the father of young children at the time. Weiner, about the same age, was newly married, and his Israeli-born wife Mira had recently given birth to their first child. Whatever the reason, the patient's MS moved him more than the other diseases he studied and treated during his long apprenticeship in medicine.

Howard Weiner was born on Christmas Day, 1944. World War II, which was still raging in Europe, had already had a devastating effect on his family by then, and his destiny was powerfully shaped by the suffering his parents had endured. His father, Paul Weiner, had been a daring young man in Vienna during the Nazi era, the youngest in a family that owned a major department store in the heart of the city. Paul was the devil-may-care son who preferred roaring around Vienna on a motorcycle, complete with sidecar, to initiation into the family business. When Howard's mother, Lola Wasserstrom, met Paul at a party in 1938, he and his pals had had too much to drink. And when he pursued her afterward, she turned him down.

But somewhere along the way she found out that the wild youth had another side: He was working for a remarkable Dutchman in Vienna named Gildemeister who was running a secret organization to get Jews out to safety. Paul Weiner was particularly involved with getting children out, and was almost caught escorting a trainload of children across the border to Switzerland.

In 1939, when the Nazi threat escalated and Lola's father, Samuel Wasserstrom, who was a furrier in Vienna, needed to get out of the country, she took the card Paul Weiner had given her, and used it to gain access to Gildemeister's operation. As a result, Lola's father received a visa to go to San Remo, an Italian town the organization used as a destination for escaping Jews. From there, he was able to travel on to Paris, where there was another branch of the family fur business. After that, Lola began to feel differently about Paul.

But very soon, survival took precedence over romance. Lola and her sister Gertie were able to leave Vienna in August 1939 for the United States, because of a family connection. Paul had no family in the United States, but he was determined to follow Lola. So he used the tricks he had learned working with Gildemeister. By the time Lola and her sister got to Denver, where relatives lived, Paul was already cooling his heels in Cleveland. He got on a bus and traveled five days and nights to join her in Colorado. Lola's father, though still in Paris, continued to exert a great influence. And he was, as Lola still remembers, "crazy about Paul." With his endorsement, the couple were married. She was nineteen and he was twenty-four.

Soon after they married, Paul Weiner joined up, first serving with the Tenth Mountain Division, the ski troops, and training in the Rockies near Denver, then returning to Europe, where he worked as a translator in the interrogation of German prisoners of war. The return home to Denver, after the war was over, required another kind of courage. He and Lola had left Europe with nothing but their passports and $4 apiece. Now they had to find a way to survive, working at jobs that they were not brought up to do.

Lola pretended she knew how to sew furs and got a job stitching together pelts with a local furrier. Paul first delivered Meadowgold milk, then moved on to selling dresses. Later he ran a laundry, then a bar. There wasn't much money, but before long the Weiners were able to move from the apartment house under the viaduct on Hooker Street, where they rented along with other immigrant families, to a miniature brick house in a Jewish neighborhood on Denver's west side.

Howard Weiner was one year old when his father returned from Europe. Like most children, he didn't give a lot of thought to his parents' economic struggle. He remembers running in and out among racks of dresses at the factory where his father worked, and earning extra money later on by stamping "White Star Linen" on sheets and pillowcases at the laundry. And he remembers visiting the fancy houses of other members of his parents' synagogue. If he really wanted to do something, his parents found a way. Howard has fond memories of going with his father to rent skis at the beginning of the ski season, and taking the ski train up to Winter Park. At the time he didn't question why his father, an expert skier, never came along. Only later did he realize that it had to do with money.

This was the way his parents wanted it. "We made sure," as his mother Lola puts it, "that whatever we couldn't do was for him to have. Because we got denied all that. We always wanted to give him everything."

Lola Weiner is sitting next to her sister Gertie in one of the booths at The Bagel, a delicatessen that is now the hub of Weiner family life in Denver. In 1969, after both Howard and his younger sister Rhoda were out of the house, Lola and Paul Weiner heard of a delicatessen that was for sale and decided to buy it. That first deli, Bagel north, has since been sold, but Bagel south is still thriving. Nowadays, it's run by Rhoda and her husband, with Lola helping out at the cash register on weekends, meeting and greeting the clientele.

The Bagel deli has a long glass case along a wall filled with the usual smoked fish, meats, and comfort foods. To the right of the door is a Ms. Pacman game, which Howard Weiner plays fervently, at a quarter per game, on his visits home. On the other side of a glass divider, Formica tables and booths in brown leatherette can seat up to 120 customers. On the wall opposite the deli case are Weiner memorabilia: a Tenth Mountain division license plate, a yellowed newspaper article about Howard Weiner's appointment to a chair at Harvard, snapshots of family groups skiing at Vail, and a big framed list, printed in black letters, with a heading in red that reads "Ten Ways to Live Longer." The advice is unsurprising: relax, find a hobby, control your emotions, see your doctor. Number ten on the list is "Eat at the BAGEL DELICATESSEN."

One photograph on the wall has special power for everyone in the Weiner family. It is an old black-and-white picture, taken in Paris, of a darkly handsome man with a mustache, flanked by two young women in pretty rayon dresses and dark stockings. The young women, whose rolled hair and clothes make them look older than their years, are Lola Weiner and her sister Gertrude. And the man between them is their father, Samuel Wasserstrom. It is his story that has haunted the family ever since they arrived in Denver.

The sisters, now in their seventies, retain some of the allure they had in the fifty-year-old photo. Their faces are lined and spotted, but their brown eyes still radiate warmth and mischief. Gertrude is blunt and to the point, and her voice has a whisky rasp. Lola is seductive, with a voice and accent that make her sound a bit like Marlene Dietrich. Together they tell the story of what happened to their father.

Because they were born in Vienna, the sisters had an easier time getting into the United States than their mother, who was on another list with a different quota because she was born in Czechoslovakia. Finally, eight months after they arrived, their mother was able to follow. But their father was still in Paris. Everyone had assumed he would be safe there. "Whoever thought that Hitler would take over France?" Lola asks rhetorically. "The Maginot line was something that nobody could touch. But it didn't take him half a day to go over it." After that the family's energy was trained on getting the father out.

"The family here got a special visa from Washington, D.C., for my dad," Gertie explains, "and a passport. He packed his suitcases to go on the train." The train went to Lisbon, where he was to proceed by ship to the United States. But somewhere along the way, perhaps at the Paris train station, perhaps during the train ride, perhaps when he was getting on the ship, someone stole his visa. Despite their best efforts, the family has never been able to find out who did it and what happened afterward. All they know for sure is that Lola's father's luggage arrived in Denver and he didn't. After the war, they learned from a survivor that their father was incinerated at Auschwitz.

The picture at The Bagel deli of the young man with his two daughters was taken three years before he died.

"He was very, very artistic," Gertie says. "He would make patterns for coats, he was a designer. Mink, Persian lamb, sable."

"Everything was made to order when you came into our place," Lola remembers.

One of Samuel Wasserstrom's dreams was to have a doctor in the family. But the idea that one of his daughters could be a doctor would never have occurred to him, the sisters say.

So when Howard was born, Lola decided he would be a doctor for her father. "I said to him, just like I'm sitting here, 'Honey, you're going to be a doctor because your Zayde [grandfather] wanted you to be a doctor.' It was instilled in him."

"She talked it into him," Gertie says, "from the day he was born. I didn't do that to my children."

Lola responds to her sister: "I did well by saying it, didn't I?"

Howard Weiner, the chosen grandson, can't remember a time before he decided to become a doctor. "As I was growing up, if you were to ask me, I would probably say I wanted to be a doctor. . . . And if I were to see doctors walking outside a hospital, I would always feel a twinge that I belonged there."

Weiner was only twenty when he started medical school, after completing three years at Dartmouth. He went to the University of Colorado at Denver because it allowed him to return to his basement room in the family house on Wolff Street and save a lot of money on board and tuition. Even though he was one of the two youngest in his class, he started off with huge ambitions. "I've made up my mind I want to be one of the top students in my class," he wrote in his journal. "I think I can do it and certainly must try. I've labeled my goal the 'uno-primo' project. Certainly I'm not going to kill myself if I'm number thirty-five, but somebody has to be number one, and why not me?"

Very quickly, though, Weiner discovered what many medical students had found out before him. Although being a doctor may be exciting, being a medical student is tedious in the extreme, and involves a great deal of rote learning that strikes many students as useless and irrelevant. By the end of the first semester, he wrote in his journal that "I was enthusiastic for school and to do well, but this was dampened as the semester went on." By the second year, he was skipping some classes because they were boring, and labeling others as "unbelievably bad," and all about "things we'll never have to know." The idea of being number one got dropped along the way.

Weiner found his classmates almost equally disappointing. "Friendships among classmates have been too few and too shallow," he complained. With only a few exceptions, his fellow medical students seemed "one-dimensional and gossipy." He was critical of the fact that so many of them seemed to be "living to satisfy others," rather than themselves, and predicted they would eventually come to a point where "there is no longer a self to satisfy." This was an intriguing criticism, coming as it did from a young man who had gone into medicine, at least in part, to satisfy his mother.

What kept Howard Weiner going during these years were his extracurricular projects. He made and edited music videos, and showed them in his own small film festival one night; he wrote songs, and he wrote poetry, sometimes even during boring lectures. He also wrote in his journal, "capturing a moment in time," as he put it, "for the purpose of nostalgia and immortality."

Without doubt, the most important extracurricular project of Weiner's medical school years had to do with women. At Dartmouth, where girls appeared only on weekends, dating was a kind of competitive game most of the time. But at medical school, he began to consider the possibility of "one special girl." "What kind of girl I want is a complex question . . . but definitely one that serves as a catalyst for me and one to whom I'll serve as a catalyst; a girl I'd want to write a poem for . . . a happy girl with a zest for life and a feeling for the humorous, the light side of things; a girl who can think." He also wrote, in Hebrew, that he wanted to find a Jewish girl.

Not long after that, Howard Weiner met a beautiful young Israeli with dark, curly hair and sparkling eyes named Mira. The story is well-worn by now: He went to a play with a nursing student he knew. During intermission, they ran into someone Howard knew from a summer job he'd had at a clothing store in Denver called Bond's. The two were exchanging news when the fellow salesman's date returned from the ladies' room. "This is Mira," his acquaintance said, "and she comes from Israel." Howard wrote in his journal at the time, "I couldn't believe my eyes."

Howard and Mira struck up a conversation in Hebrew, while their dates looked on. During the second act, Howard couldn't concentrate on the play. Afterward, when the two couples went out for a snack at the Red Slipper, he sat across from Mira and learned more. She was from Ramat Gan, had already served two years in the Israeli army, was twenty-one years old like him, was studying at Denver University, and lived with an uncle in Denver. The next day, he called Bond's and talked to another salesman there, to find out if there was anything serious between Mira and her date of the previous evening. When he learned there wasn't, he called her up and asked her out to a Bennet Cerf lecture. "And so," he wrote in his journal, "I met a Jewish girl. And since I wrote my wish in Hebrew, she is Israeli."

From the start, Mira was sympathetic to Howard's very lofty aspirations. "I decided," she says now, "that I would never do anything to clip his wings." When they saw *Dr. Zhivago* together, and he identified with the doctor-poet at the center of the story, she seemed to understand what he felt. "I put my head next to hers and we shared the same emotion—simply wonderful." At the same time, Mira was no pushover. She reminded him that he frequently acted spoiled and selfish. After his mother told him that she had "lost a son," now that he was in love, Mira ordered him to be nicer to his mother. And sometimes, when he was grandiose, she burst his bubble. Once he showed her some drawings he had done. They were terrible, he knew, but he told her they would be valuable one day when he was famous. Mira asked, deadpan, "How long do we have to wait until you become famous?"

Medical school improved toward the end of the second year, around the same time that Howard and Mira decided to get married. For the first time, there were real patients to examine and learn from. Howard wrote in his journal that physical diagnosis was "exciting." He became especially intrigued with the brain. "I'm fascinated by the creative process," he wrote in his journal. "What goes on when man creates? To me this is the key to man's superiority. Are all his thoughts, his loves, his imaginations only chemistry, a series of reactions? I think so. But the formulas and reactions are so numerous that they defy understanding." That spring, he did a neurological exam on a patient at the VA hospital and described it in his journal in capital letters as "EXCELLENT!" He added, "at times I am thinking of going into neurology, and am fascinated by the mystery of multiple sclerosis."

During his residency, when Weiner decided to pursue the mystery, he sought out the only person in the area focusing on MS, a neurologist at Massachusetts General Hospital named Barry Arnason, and worked for two summers in his laboratory. There he encountered the sine qua non of research on MS, a disease called experimental autoimmune encephalomyelitis (EAE), which can be induced in mice and rats. This murine equivalent of MS was, and remains, the best way to study the disease in a live creature. It was with Arnason that Weiner attended the first of many hundreds of conferences. At a meeting in Montreal, he watched in awe as his mentor held forth on MS and the immune system before a large audience.

At the time, there were two major themes under discussion among MS researchers. One was the possibility that MS was triggered by a virus. This was an appealing idea: If a viral culprit could be identified, MS could perhaps be treated. The fact that a number of other neurological diseases, including kuru and Creutzfeldt-Jakob disease, were thought to be caused by viruses made the hypothesis seem more likely.

There were other reasons to suspect a virus. Although there is some genetic predisposition to the disease, it is minor compared to the curious role of environment. MS is a disease of northern latitudes, common in Ireland and Scotland, less common in the Middle East. It is almost nonexistent among African blacks, yet common among Scandinavians. A study of northern Europeans who moved to South Africa at a young age showed they tended to be susceptible at the lower rate of their South African compatriots. Something decisive was happening during the early years of life. But what? It was natural to speculate that some sort of virus, contracted in childhood, was the trigger of MS.

The second focus among MS researchers at the time was in the murkier area of the immune system. Whatever the trigger, it was clear that the body's own lymphocytes were going on the attack and causing inflammation. Why did the immune system, so necessary for protection against disease, run amuck in multiple sclerosis?

After his neurology residency in Boston ended in 1974, Weiner chose to return home to Denver, Colorado, for a postdoctoral fellowship in immunology that would provide the basis for answering that question. Under Henry Claman, a pioneer in distinguishing B cells from T cells, Weiner began to learn "the rigors of cellular immunology." There were many hours spent at the bench with John Moorhead, then a junior faculty member, doing the repetitive work of centrifuging cells and measuring proliferation. But he also did some studies related to MS. "They were kind of journeyman studies," he says. "But I was asking MS questions, which is what I wanted to do."

As a postdoc in Denver, Weiner initiated the double life that he has lived ever since, moving back and forth between the clinic and the lab. Money was tight—he had two young children by then and was earning $12,000 on his fellowship—so he moonlighted on Friday nights in the intensive care unit at St. Anthony's Hospital. During the week, he worked only with mice. On the weekend, he treated humans in extremis, some of whom arrived by helicopter on the hospital roof.

Weiner had always been close to his family, and considered staying in Denver. But his wife Mira, in retrospect, is glad they chose to return to Boston. "None of our life could have turned out the way it did if we were in Denver," Mira says. In the first place, there was no obvious way for Weiner to advance in Denver. Second, Weiner's family, and particularly the well-meaning but relentless involvement of his mother Lola, would have made it difficult for the young couple to establish a separate existence.

Back in Boston in 1976, the Weiners settled into a modest house in Brookline with their two young sons. Howard, now in his early thirties, took a second fellowship, this time in the lab of a renowned virologist named Bernie Fields. In one way, it was a detour from MS, since Fields was working on reovirus, an animal virus with no human equivalent. But it was in Fields's lab that Weiner learned the most about how to be a lab director and mentor.

Weiner did his first significant work in the Fields lab, isolating a gene that made the difference between life and death in reovirus-infected mice. "Some of the biology we look at is very subtle, it's a little band on a gel or it's a number. But the readout in this experiment was life or death. You inoculated them as infants and then you'd come in the morning and see how many were alive." Weiner placed some of the mouse cages on his office desk so he could keep constant watch. "You couldn't do that now," he notes, "but you could in 1974."

Weiner and Fields had a close relationship. "He was very much a father, mentor, friend," Weiner remembers. "He did science at the highest level. It's sort of like listening to one of the best violinists so you know what it sounds like when someone really good plays. If you've never heard it, you don't understand what it's like at that level."

It helped that Fields's work had a theme, that he wasn't tied to convention, and that he was good with people. But some of it was mysterious: "Things always seemed to be happening good to him in the lab. It just seemed to be working. There was a certain rhythm, and a certain way things came together. I can't explain it exactly." Sometimes, when things are going well in his own lab, Weiner gets "that same feeling."

At the same time, there was trouble in this paradise—the trouble that arises almost inevitably between a lab director and an ambitious postdoc. Weiner felt confined inside Bernie Fields's elaborate system for studying reovirus. "The reovirus," Weiner explains, "hasn't had any major applications. It's really a model virus." And he was beginning to sense a major difference between him and his mentor. "Bernie liked to fiddle in the lab. Bernie liked the models. And in fact he gave up seeing patients in the hospital very early in his career. When I was there he hung up his white coat for good."

Weiner, on the other hand, had always wanted to do work that would have clinical applications. As he began to build up his MS clinic and take his work in an independent direction, the tension between him and Bernie Fields mounted. Fields believed Weiner hadn't given him enough credit in one of his presentations. Weiner, for his part, was hurt that Fields didn't acknowledge him, and resented the fact that another postdoc, who succeeded him, was taking credit for minor variations on experiments he had already done.

It wasn't until Bernie Fields entered Weiner's regular poker game, many years later, that the two reestablished their friendship. After Fields contracted pancreatic cancer, Weiner took part in the public tributes to his mentor, who in turn acknowledged the important contributions Weiner had made in his laboratory. Near the end, Weiner visited Bernie Fields in the hospital and kissed him goodbye. But even after the breach was healed, Weiner never forgot, as director, the hurt he had felt. He resolved not to inflict it on those who worked under him. It turned out to be a hard resolution to keep.

—◦—

The Center for Neurological Diseases, established in 1985, was the brainchild of Weiner and Dennis Selkoe, a neurologist who has done pioneering work on Alzheimer's disease. Weiner was one year ahead of Selkoe in residency, and showed him around when he first arrived in Boston with his wife, back in 1971. "Howard helped me learn the ropes," Selkoe recalls. "Where to live, where the stores are." When Dennis Selkoe says "Howard," the word is laden with his long acquaintance. "Howard was always like he is now," Selkoe says. "You could use the word proactive or you could use the word aggressive. He was just very much involved in everything, loved to be in the action, in the center of things."

As soon as he began his residency, Selkoe got a taste of Weiner's enterprising ways. Weiner and his senior resident, Larry Levitt, put together a series of practical mimeographed notes that began with the patient's presenting symptoms and proceeded step by step to diagnosis and treatment. When everyone started making copies, Weiner and Levitt decided to turn their handouts into a book. In the ensuing years *Neurology for the House Officer* became the most popular neurology manual in the country. It now sells about 10,000 copies a year, and has been translated into several foreign languages. It has also been used as a model for an entire "house officer" series, written by other specialists with a credit to Levitt and Weiner in the back.

Selkoe also remembers Weiner's teaching style as chief resident. He would show an X ray, or an electroencephalogram (EEG) script without explanation. Then at the weekly lunch, he would ask the residents to make their guesses about what they were looking at. "That was very typical of Howard," Selkoe says. "He's very competitive and likes to see who does better." It is also true, as all who know him will attest, that Howard Weiner likes to turn everything into a game.

For years, whenever they met, Selkoe would encounter the same competitive streak. "He was always the kind of guy who would ask a lot of questions—where do you stand, how are your grants going? Once in a while it bugged me. Of course he didn't care about economic status or social status. He wanted to know, in what we both are working on, where are you versus where am I?"

A black-and-white picture of Weiner and Selkoe, taken in the hospital ward in 1972, now hangs in the offices of the center. Weiner, the taller of the two, appears even larger because he is in the foreground. His flying hair makes him look like a mad scientist, pen poised above a white pad. Selkoe, slighter, hangs back like the junior resident he is, leaning on a counter. He also has longish hair and sideburns, de rigueur in the seventies, but it is carefully parted and slicked down. He looks as though he is amused and perhaps slightly uncomfortable with his friend Howard's theatrics.

In the years since the photograph was taken, Selkoe has come into his own. His work on Alzheimer's has contributed to the understanding of the underlying causes of the disease and to the founding of a company that is about to initiate the first human trials of a potential vaccine. He is a man whose looks have improved with age and confidence—a very slim, fit man in his early fifties with Wedgwood blue eyes and a direct gaze. Unlike Weiner, who sometimes has a shirttail hanging out, Selkoe is fastidious, and apologizes for the dead leaves a plant has shed on his immaculate office carpet.

The idea for the center came to Selkoe and Weiner in 1985, after they wound up with labs on the same floor of one of the buildings in the Harvard hospitals complex. If they had a center, they reasoned, they would be better able to raise money. They could provide their researchers with an administrative structure and with the stability that they needed when grant money flagged. The center had its critics in the hospital leadership; there were some who thought it pretentious and unnecessary. But it has turned out to be a brilliant stroke. Although they have had serious conflicts during their fourteen years of working together, Dennis Selkoe and Howard Weiner have, in Weiner's words, "a marriage that works."

The children of this "marriage" are the young men and women who come to the center as postdoctoral fellows and who do the work at the bench, day in and day out. They are heavily educated, sometimes with M.D.'s as well as Ph.D.'s on their résumés. Yet their pay is modest—starting at about $25,000 a year in the Weiner lab—and most of them work ten-hour days, six days a week. They come with ideas of their own, but must pursue the themes of lab directors who no longer touch a mouse or pick up a pipette themselves. In the Weiner lab, most of the postdocs come from other countries. Americans usually aren't willing to tolerate the modest lifestyle or the long hours, and instead find work in the more lucrative pharmaceutical industry.

Yet those who do come as postdocs marvel at the riches of the center, compared with what they've had in their labs in other countries. Anthony Slavin, who came to the Weiner lab from Australia, says he can have an idea for an experiment at the center, make three or four phone calls, and have all the materials he needs to do it the next day. "In Australia I'd have to plan for three weeks." For better or worse, he says, the American approach is "let's just get a whole lot of good people and let's give them free rein and get them to do as many experiments as they can and sooner or later they'll find something."

There is a cost for such riches, according to a postdoc from South America. "Back there, we feel we probably aren't going to make a big difference in the world, whereas at Harvard you feel you are on top of the mountain. But it isn't as warm and friendly."

Not surprisingly, postdocs talk a lot about lab directors, good and bad. Everyone has at least one shocking story of abuse. Most egregious is the story of the Japanese Nobel Prize winner who calls his postdocs by number rather than name because it's less bother. In some labs, the director deliberately gives two postdocs the same assignment, setting them up for a race to the finish. The result is that the two don't speak to each other, let alone share findings.

"People are very harsh," notes Youhai Chen, who has risen from a postdoc position in Weiner's lab to being an associate professor in charge of his own lab at the University of Pennsylvania. "Everybody is critical of themselves and of others, and that can create barriers." Often, now that he runs his own smaller lab, Chen thinks about Weiner's approach. "It's not easy running a big lab like that. He's one of the best, very supportive, and he thinks about you a lot. He is concerned about direction and the big questions. He helps you release your energy."

At weekly lab meetings, Weiner always sits right up next to the presenter, and is one of the first to jump up and fiddle with the overhead projector when it breaks down or to lend his laser pointer. His comments are often designed to put people at ease and deflect criticism. Sometimes they are so general and simple-minded as to elicit titters from the sophisticated scientists in the room. At the beginning of a presentation on a hunt for an MS gene, for instance, he notes that "people often don't understand genetics." Or he will ask, "Does everybody understand this?" He asks one postdoc what M cells do in the gut, giving the postdoc an opportunity to hit a home run. Another time, he introduces a post-doc by announcing that the work he is presenting will be coming out in a month in a good journal, after a range of reactions from peer reviewers who saw it as terrible, nothing new, and "the greatest thing since pasteurized milk." Everyone in the room empathizes—they've been there. After the presentation, he restates and anoints questions he likes. "Jeffrey's question is an interesting one," he will say. When he wants the postdoc to do something more, he doesn't issue a direct order. Rather, he makes a suggestion, squinting and gazing into the distance as he speaks.

If it were possible, Weiner would prefer not to issue orders or make unilateral decisions at all. He hates the inevitable arguments that come up over who will get top billing and appear as first author on a paper. "It's better if you decide ahead of time," he notes, but "sometimes you just have to flip a coin." Weiner's fault as a lab director may be that he avoids making tough decisions. "Howard never says no," a close associate notes. Another, thinking about this, adds that if Howard never says no, he also never says yes. He'd rather not be the one to cut the Gordian knot. Recently, when two researchers came up with contradictory results, he invited an eminent outsider to come in and discuss the same issue. That created a situation in which two were on one side and one on the other. The researcher in the minority, many noted, was "hung out to dry" and Weiner emerged unscathed—still the good guy.

Sooner or later, if they have accomplished good work, postdoctoral fellows want to move up and direct their own research, just as Howard Weiner did. Tensions arise. But Weiner's goal has been to stay connected, to avoid "getting into a thing where you can't come back together." Always, in the back of his mind, is his memory of "what happened with me and Bernie."

"It's easier if they leave," Weiner acknowledges, and cites a string of researchers, like Youhai Chen, who have gone from his lab to leadership positions elsewhere. Of those who have stayed, the one who has presented the greatest challenge is without a doubt one of Weiner's earliest postdocs, David Hafler.

Hafler has a plump face, dark curly hair, the eagerness of a preadolescent, and, sometimes, about as much tact. He once brought leftovers to work from a restaurant meal the night before and asked the owner of the lunch place to warm them up for him. He will tell you that he had his first microscope at five, and became interested in immunology at seven. Before he finished high school, he was doing a project in immunology at the university. Eight years younger than Weiner, Hafler has stayed on to become a principal investigator. Along the way, he and his former mentor have struggled to establish a modus operandi. Hafler is sometimes competitive with Weiner and at other times deferential. He can be seen standing on the edge of a lab meeting, for instance, mumbling loudly that the work of one of Weiner's postdocs is "not ready for prime time." But then, when he has to leave another lab meeting early, Hafler makes an excuse to Weiner, as though he were a schoolboy needing permission to leave class. The two of them describe their difficulties over the years in very similar terms. "It's a classic conflict that occurs in science," Hafler notes. "What do they take with them, what's theirs and what's yours." The issue, in Weiner's view, has been "where's his place and where's my place?"

"There was a great deal of friction between Howard and me as I started to become more senior," Hafler says. "Each step along the way. Publishing as senior author with Howard on the paper. Then publishing without Howard on the paper." "We've had problems," Weiner acknowledges, "but we've always resolved them. That doesn't mean I didn't sometimes want to tear his hair out. And I'm sure he wanted to tear out mine too. Sometimes we had long sessions. Once there was a problem where I said, 'David, this is going to take four sessions. So let's start, let's do session one.'"

Recently, David Hafler was named the Breakstone Professor of Neurology and Neuroscience at Harvard. Howard Weiner spoke at the ceremony where he was given the chair, acknowledging his important contribution to the study of MS. The professorship and the chair, Weiner noted, are tributes to Hafler's important scientific contribution and "to the ability of people to grow up within the Center."

"The amazing thing," Weiner now says, "is that we made it through."

CHAPTER 4

The Researchers

IT'S RAINING OUTSIDE THE OLD BOSTON LYING-IN BUILDING, now the Center for Neurological Diseases. The granite stones of the graceful curved driveway are slippery, and the swaddled babies on the entryway frieze are soaked. On Longwood Avenue, a few yards from the entrance, the noises of the vast complex of Harvard hospitals regularly disturb the watery air. The yodel of sirens is punctuated by the roar of the medivac helicopter, whose echoes ricochet off the buildings as it descends onto the pad on top of Brigham and Women's. But inside the center, all is quiet. Umbrellas, left open to dry, make a gay show of color along one side of the long hall outside the laboratory. Posters describing current projects, complete with pictures and multisyllabic headlines in large letters, line one wall, opposite lockers with wooden doors. Occasionally a worker emerges from the lab and hurries out the door to the animal room with an armful of vials.

The lab itself is a sort of warren—a series of interconnected rooms off the long hallway, all intersected with row after row of waist-high slate-topped benches. Researchers move silently about, accompanied by the soft swooshes, beeps, and vibrations of machines. Or they talk quietly as they graph their results on computers or dissect mice side by side under the intimate glow of the hoods.

Sometimes they talk of science. Other times they talk of mundane things: a Filene's basement sale, plans for the weekend, picking up a car from the mechanic. One of the Japanese postdocs has just fathered a premature baby, and all who have seen it are amazed at how tiny the baby is. Australian postdoc Anthony Slavin talks of plans to travel to Africa during an imminent vacation. They are young people, for the most part, and far from home.

In a recent tally, the 140 people working at the Center for Neurological Diseases came from thirty-three different countries. Everyone is supposed to speak English in the lab, so no one is excluded, but it's not unusual for two Japanese postdocs to break the rule. Even the Israelis, more practiced in English, sometimes shift into Hebrew among themselves.

The lab has a complex geography. Each principal investigator has an office that is the center of a fiefdom, where varying numbers of postdocs and technicians vigorously probe one small area of autoimmune disease. Farthest from Weiner's office are the groups who are studying human blood and tissue, including David Hafler's group. Nearest to him are those whose research involves animals, including his own team. The "human people" and the "animal people" don't always see things in the same way. "The human people don't like to rely on our results," says Ruth Maron, "but sometimes they have no choice, because often we have the only data."

Ruth Maron is an anomaly in the Weiner lab, a researcher in her forties who is neither a principal investigator nor a postdoc. Although she has the credentials and experience to run her own lab, she perceived early on that such a position would force her to spend a lot of time generating money, and pull her away from the benchwork she loves. Four years ago, she took a demotion in rank from assistant professor at Tufts to instructor at Harvard so she could join the Weiner lab. "You have to be working on your own grants here to be an assistant professor," she explains. "I like to put my hands in the water—it's more important to me than the title."

Maron wears no makeup, and her gray hair is pulled back into a thin ponytail. She saves the color for her wardrobe, including a seemingly inexhaustible repertoire of shoes and boots in vivid colors. Born in Germany just after the war, Maron grew up in Israel in a family scarred by the Holocaust: Her parents had lost two children before she came along. She too has had losses: A first husband died young, leaving her to raise a son on her own. She never fulfilled her childhood dream of becoming a doctor. Instead, she has pursued a career in science, earning a Ph.D. while working in cellular immunology at the Weizman Institute in Israel. It was there that she began to focus on autoimmune disease. During a postdoctoral fellowship in Boston she met a new man and began a new family life.

The scientific adjustment to America took longer. At the Weizman Institute, she had been "queen," whereas at the Joslin Clinic in Boston, where she began as a postdoc, no one cared much about her. "In Israel," she explains, "you have someone who chaperones you, teaches you, shows you what's going on. Here it's jump into the pool and if you can swim you survive, if you can't you drown."

Ruth met Howard Weiner when she came seeking advice about a friend with multiple sclerosis. She was working in a lab at Tufts at the time, and felt there was "no one to talk to" there and very little excitement about the work she was doing. Weiner encouraged her to come back when she was ready for a change and talk to him.

"I think we sort of found each other," Ruth says of her relationship with Howard. "After I came here, I learned that a lot of people were advising Howard that he needed someone to be around the lab for people to talk to, because he's very busy." "Ruthie," as she is universally known around the lab, has filled some of the gaps.

Ruthie is also the person Weiner comes to when he wants to try something new. "I'm the rabbit," she says. "If there's a new idea he wants to introduce into the lab, he asks me to try it out." Successful ideas often are passed on to other researchers, but Ruth is willing to accept that price in exchange for the excitement. Even though she's working harder than she would like, she loves the atmosphere in the lab. "I like it that there are people you can talk to, that everyone is really interested in what you are doing and has the same aim. I'm very happy every morning when I go to work."

On this particular morning, Ruthie's role as nurturer has taken over, and she is at her desk in the lab trying to make arrangements in several languages. There is a phone call from the school nurse about her eleven-year-old daughter, who is feeling sick. And there is a conversation in Yiddish with her mother, who lives with her, followed by a worried conversation in Polish with her mother's housekeeper. Her mother is ill just now, and tears well up when she talks about it. Finally, there is a conversation with her technician, Nancy Melican, who is home sick with a bad virus.

"And don't come in tomorrow if you're not feeling well," Ruth instructs over the phone. "If you come in tomorrow, you won't be able to come in the next day, so what's the difference?" Ruth's voice rises in a combination of sympathy and indignation, and her thick Israeli-accented "r's" stretch out words like "tomorrow" and "difference" to formidable length. Sometimes newcomers who overhear her on the phone think she's angry when she's just making her point—forcibly.

With Nancy out sick, Ruth must attend to the animal experiments she has under way. She begins by washing a little syringe repeatedly, pulling water in and out. Then she gathers small vials of substances from the refrigerator and puts them all in a plastic container, throws on her down jacket, heads down the hall, and swings open the door into the cold wet November air.

Ruth is used to tending the animal experiments: She does it on Saturdays and Sundays because it's a long trip into the labs for Nancy. Yet even when the mice are producing great results, she is still troubled after all these years by the animal work. "I find it stressful," she says. "I don't like treating living beings as test tubes." With a sad smile, she tells of her son's announcement that he was going into computers instead of biology because "you've already killed enough animals."

Ruth and others use the word "sacrifice" to describe the killing of animals in their work. It is a euphemism that contains a justification: The animals are sacrificed for the good of humankind. What makes animal research acceptable to Ruth and the others involved in it are the results: Animal research has produced cures for many scourges, including diphtheria, tuberculosis, and polio. Almost everything that is now known about autoimmune disease has come from animal studies.

There is a paradox, however, at the center of animal research: It is because the bodies, the brains, and sometimes even the psyches of animals are like ours that they make good models for the study of human ailments. And it is because the animals are like us that we have qualms about using them in this way.

Fifty yards or so on the other side of Longwood Avenue, Ruth enters another world: the animal room in the Warren Alpert building, a cool gray granite addition fused onto one wing of the Harvard Medical School. Security is tight here, but Ruth nods to the guard, who knows her well, and swipes her plastic card to unlock the door to the lab. The protests of animal rights activists have made such precautions necessary. Also, partly because of pressure from animal rights groups, the cages are large and well maintained. Each of the thousands of mice in this facility was purchased for around $15 and is maintained at about thirty-three cents a day. Ruth describes the facility, in fact, as a "five-star hotel."

Yet it is definitely a rodent hotel. From the moment Ruth passes through the secured door, the animal smell is thick in the air. In a dressing room, Ruth strips to her underwear and pulls on blue surgical scrubs over her slim figure and slips disposable booties over her shoes and a disposable cap over her ponytail. All of this is part of the effort to keep pathogens away from the animals.

In the control room, the attendant who monitors entries and exits from behind glass tells Ruth that there was a flood in one of her cages and the animals died. "I don't think it was one of the critical experiments," she says, "but we'll see." Using her card one more time, Ruth opens the door into a room of stacked cages filled with hundreds of mice. There are many kinds of mice here, some black, some white, some with large litters of tiny blind pups, others with older babies nursing at their mother's teats. Each cage has a clean lining of cedar chips, and each has a card taped to it that tells a story: what breed the mice are, who the researcher is, and what's being done to them. Ruth sets down her supplies, pulls on thin gloves, collects her thoughts, and reaches for a cage to begin her work.

As usual, Ruth has several experiments going at once, because, as she explains, "if you only do one thing, there aren't enough hours in the day." Some of the experiments are on familiar territory with what she calls "my diabetic mice," a breed of mice who naturally contract diabetes. Another experiment—the one she's most excited about at the moment—is exploring what may be common ground between two different diseases: EAE (mouse MS) and diabetes. Anthony Slavin, the Australian postdoc in the lab, has discovered that her diabetic mice are capable of contracting EAE if given a certain protein. Today she's looking for confirmation of this finding.

Swiftly, Ruth reaches into the cage and grabs the pink tail of one of the little white mice. Then, with her other hand, she takes hold of it firmly by the neck. She wants to see if the mice in this group are showing signs of EAE. The tail is a good indicator of disease, since paralysis sets in there first. Some of the mouse tails curl actively around Ruth's finger, indicating that they are healthy. But other tails are listless, and some of the mice are dragging their whole rear ends behind them, using their front feet. Ruth observes and assesses each mouse, using a scale in which zero means no disease and five means death. She looks at their fur, which turns scruffy and loses its lustrous white with disease. And she turns them over on their backs to see how hard it is for the little creatures to right themselves. There are some who have become sicker, and others are getting better: In short, the mice have ups and downs similar to those experienced by human patients with relapsing or remitting multiple sclerosis. Before she leaves, Ruth sprays water on the food pellets of the afflicted mice, because some of them have a hard time getting to their water bottle.

In the animal room, Ruth works alone. But back at the lab, she almost always has company, usually the two postdocs she is working closest with at present, Anthony Slavin and Gabriela Garcia, who comes from Brazil. She tells them the good news that the diabetic mice seem to be contracting EAE. And she tells them about the flooding, which fortunately didn't wreck any experiments. Conversation then turns, as often happens, to mice in general. Anthony's progress, it seems, has been slowed by a major cell death, which has forced him to start over breeding a new batch of mice. Ruth remembers that when she was finishing her Ph.D., "all the mice got some terrible disease and had to be sacrificed." She was devastated, but fortunately she had an easygoing adviser who told her, "Well, this way you finished your Ph.D. sooner!"

Everyone who works with animals has stories about how this illness or that one wiped out months of research. Whole animal labs have been rendered useless by one virus or another. Right now there is a worry about the animal room in the Alpert building. Some of the mice's eyes are looking red and the vet, who keeps a close watch, thinks there may be conjunctivitis going around.

The kind of genetically engineered mice these researchers use are hard to replace. Bred through sophisticated processes on experimental mouse farms, they aren't always available in the desired numbers. The Center for Neurological Diseases can rarely fill its standing order. Yet without them, the work can't go forward. In the hierarchy of the lab, the mice may be at the bottom. But like most underlings, they are what keeps the whole operation going. "The mice," Ruth Maron says, laughing, "are more important than we are."

"Until four years ago," Anthony notes, "we couldn't do an MRI [magnetic resonance imaging]. The only way to see the effect of MS in humans was on an autopsy." The animal model for MS, EAE, was and still is essential to studying the disease's effect. Yet EAE is just a model, which means it can tell researchers only so much about human disease. Anthony once watched a lecturer make this point in dramatic fashion. On the first slide, he enumerated the compounds that have been discovered that cure mice of EAE. It was a fairly long list. On the second slide, he listed the compounds that have been discovered that cure humans of multiple sclerosis. The second slide was blank.

As the researchers talk, their ambivalence toward the mice emerges: They need them, but they know they can be let down by them. Genetically identical animals allow scientists to control variables to come up with meaningful results. But the genetic tricks create animals that are less like humans than wild ones would be. And then they use these genetically engineered animals to try to understand the diseases of humans, who are genetically unique and various. It would be easier, someone suggests jokingly, if humans could be genetically engineered. "I'm glad they're not," Gabriela Garcia immediately replies.

When Gabriela came to the lab two and a half years earlier to study rheumatoid arthritis, her first challenge was with the animals. "I needed to get 100 percent sick mice," she explains, "and it was very hard to do. I would only get 70 percent or 80 percent." Another researcher who had gone before her had failed in the attempt, and she feared the same fate. But when she went to Howard Weiner with her worries, he pointed out that the previous researcher had no lab experience, whereas she did. "We play basketball here," he told her. "She didn't play basketball." To a Brazilian, the basketball metaphor was more mystifying than reassuring.

Gabriela is a petite woman in her thirties with large brown eyes and dark shiny hair. She has always had powerful mentors leading her as a scientist. In Brazil, she worked in the laboratory of Nelson Vaz, a giant in the field of immunology, and says she is "still working with him in my mind." Then, in the summer of 1992, attending her first international immunology conference in Prague, she heard Howard Weiner speak. Later, at a party in a beautiful Prague courtyard, she summoned the courage to walk over and join a conversation he was having. Weiner asked her, "Do you want to listen or do you have questions?" Gabriela, despite minimal English, told him she had lots of questions. In the end, after several more conversations, Weiner told her to get in touch when she finished her Ph.D.

Gabriela's desk at the Center for Neurological Diseases is a testimonial to all her attachments. Below the shelf of immunology texts and next to "Gabriela Garcia, Ph.D.," printed out in ornate script, is a dreamy black-and-white photo of her, with hair flowing, alongside her Brazilian mentor, Nelson Vaz. There are photos with children, photos with friends, and, despite the fact that she sees them in the flesh all the time, photos of happy occasions with Anthony and Ruthie. Amid the photographs are the lyrics to "San Diego Serenade" by Tom Waits.

Never saw my hometown until I stayed away too long
Never heard the melody until I needed the song

Leaving Brazil for Boston wasn't easy for Gabriela. Her marriage had just broken up and she was in danger of becoming homesick and depressed. But she has kept her connection with home by e-mailing her mother every morning. And she made the decision to rent an apartment she loved near the Public Garden, even though it stretched her modest fellowship budget. She also developed a passion for Filene's basement, where she regularly shops, "especially for coats." Most important, though, have been her lab relationships with Anthony and Ruth.

Howard Weiner always responded to her requests, but he was usually in a rush. "Because of the language," she says, "it takes time, you know?" The person who had the time was Ruthie. "An angel came into my life," Gabriela says of Ruth Maron. And then, with considerable exaggeration, "Ruthie taught me everything I know." Anthony has become her close collaborator and pal. He calls her "mooch"—short for *muchacha*—and the two tease and banter constantly. Even Ruthie, the serious and purposeful one, gets drawn into the play at times and can be heard laughing over at her desk.

Anthony Slavin has a big, slightly crooked face, with rosy cheeks so full they crowd his mouth and blond curls that tend to flop down over his forehead. Humor is a habit, if not a compulsion. It is rare to be in his company for more than a few minutes without a joke, even when he's in Howard Weiner's office.

Howard to Anthony: What do you think of the idea of using 1 milligram?

Anthony to Howard (poker-faced): It was your idea so I think it's a good idea.

Later in the same meeting, Howard goes into his computer in search of his notes on Anthony Slavin's research and can't find them.

Anthony asks him, "Did you look under Andrew?"

Howard comes back at him: Well, Richard, if you'd do some good science . . .

Anthony: Howard has a hard time remembering me.

Howard: We don't like Australians here.

Then, after a pause, "Actually Anthony has been a great addition to the laboratory. I say that in his presence. We appreciate him."

Anthony's face reddens and he smiles with pleasure.

What Anthony misses most, in the United States, are the friendships with his "mates" and family at home, where most people know each other for life and place more value on that deep connection than on their work. In Australia, he says, it's not unusual to talk with someone a long time at a party without finding out what they do for a living. "Here in this country, almost the first question someone asks is what your work is." When you leave home, he says, you lose a big part of who you are. "Your identity is very much the place you came from, and your family, and your friends, and you give that all up when you stay here."

And yet it was Anthony's idea to come, because of his work. In Australia, he was in the lab of Claude Bernard, a pioneer in the study of EAE in mice. But he knew, as all researchers do, that he would never have the resources in Australia that he would in the United States to pursue his work. He was especially interested in a promising new approach to autoimmune disease called oral tolerance. And there was no better place to study that than the Weiner lab. "This place," Anthony says with a lopsided grin, "is famous."

CHAPTER 5

The Big Idea

ANTHONY SLAVIN HAS AN AUSTRALIAN DISLIKE OF FORMALITY, and a particular aversion to neckties. But for his presentation of a paper to several hundred biologists at a conference in Washington D.C., he did exchange his usual knit pullover for a white shirt. He went along with another convention as well: he began his talk with the standard slide and opening statement of the Weiner lab:

> Our lab is interested in oral tolerance. Oral tolerance refers to the long-held observation of systemic hyporesponsiveness to an antigen fed prior to immunization.

Howard Weiner and five others from the lab were sitting shoulder to shoulder in the front row, in a show of solidarity. Anthony looked down toward them with just the hint of a smile as he spoke the words that have become so familiar he could repeat them in his sleep. Shortly before his talk, he and another Weiner postdoc had laughed together about this opening gambit. It has become a sort of creed, like "as it was in the beginning, is now and ever shall be." Like every other creed, it is a restatement of what everyone in the immunology community already knows to be true: The Weiner lab is committed to an unusual but promising approach to the treatment of autoimmune disease called oral tolerance.

Howard Weiner came to oral tolerance the way he has come to everything in his scientific career: through his obsession with finding a cure for multiple sclerosis. In the seventies and eighties, he tried a lot of other things, some of which had promise. There is a slide from those early years of Weiner in a reclining chair, with intravenous needles going into both arms, and tubes rising to a bag on a rack above his head. He is going through a "sham" plasma exchange, trying out the process to see how it will work in a trial of MS patients. He is flanked, in the photograph, by David Hafler on one side and another early fellow, Stephen Hauser, on the other, both wearing long white coats. They have brought him a consolation prize for his pains—a bedpan filled with red roses that rests on his belly.

Plasma exchange—the process of actually drawing the blood of MS patients and cleaning it to reduce factors in the plasma that might contribute to MS, then reintroducing it into the body—didn't turn out to be very effective for MS, though there are certain rare cases of "fulminant" MS in which it is still used as a last resort.

But it was another, more controversial trial that shaped Howard Weiner's reputation before he came up with oral tolerance. He was still working for Bernie Fields, and Steve Hauser, then a neurology resident, told him about two Europeans who had tried treating MS with a powerful chemotherapy agent and immunosuppressant called cyclophosphamide (also known by the brand name Cytoxan), commonly used in cancer regimens. Cyclophosphamide is a harsh drug whose toxic side effects include loss of hair and a compromised immune system. Nonetheless, Weiner was intrigued by the possibility that it might work in severe MS. Before long, the two young men had embarked on an experiment that would cause them no end of trouble.

"It was a one-year trial," Weiner explains. The patients had very active MS, and almost no other treatment options. "And we found results that were quite dramatic." When they were published in the *New England Journal of Medicine* in 1983, the findings created a small sensation. Weiner's picture appeared in the *New York Times* and a local news program was broadcast from his laboratory.

Over the next several years, however, the initial excitement about the breakthrough was drowned in a flood of criticism. Some who tried couldn't replicate the study. Among 168 MS patients in a subsequent two-year Canadian study, published in the *Lancet* in 1991, there was no difference between the group treated with cyclophosphamide and other groups.

Weiner's study was sharply criticized because there was no placebo group, and because the neurologists making the assessments weren't blinded. "They broke all the rules of clinical trials," according to neurologist George Ebers, a researcher in the Canadian group. Ebers was one of those who kept after Weiner about the cyclophosphamide study. He wrote a letter to *Neurology* in 1994, complaining that Weiner had been using cyclophosphamide for a decade without evidence that it worked. At a conference where they shared the podium, Ebers challenged Weiner in front of a crowd of colleagues, asking, "When are you going to learn how to conduct a clinical trial?"

The word about Weiner, in some circles, was that he was irresponsible. "Some people," according to one colleague, "thought he fabricated the results." Another colleague, who begins by insisting, "I *like* Howard," goes on to observe that "he really hurt himself *badly* with cyclophosphamide. I'm amazed he's been able to keep all the balls in the air after that."

Weiner concedes that there were flaws in his first cyclophosphamide study. It was, he says, impossible to have a placebo group, because patients receiving the drug had visible side effects, including hair loss, that would distinguish them from patients receiving a placebo. But he points out that "we were giving it when there weren't *any* drugs." The study was important, in his view, because it showed that cyclophosphamide, a suppressor of the immune system, slowed the course of the disease. This pointed to the role that the immune system played in MS.

Over the years, Weiner has continued to treat certain patients with the cyclophosphamide regimen. "It has limited but real use in the armamentarium of physicians," he says, "for the treatment of carefully selected MS patients. We've probably treated 500 to 600 patients with it, and continue to use it, although it's not the answer to the disease." Recently, studies by other researchers have confirmed the effectiveness of cyclophosphamide, particularly in the treatment of young patients with rapidly progressing disease. There is even evidence that it may reduce the size of MS lesions.

Derek Smith, a physician in Weiner's group, says that "people have sort of come full circle on the cyclophosphamide . . . it's kind of their dirty little secret. If all else fails they do use cyclophosphamide. They know that it works."

Early in 2000, an expert panel recommended that the FDA give approval to a similar cancer chemotherapy agent, mitoxantrone, as a treatment for progressive MS, following a study that demonstrated that it made a substantial difference. For Weiner, who was quoted in the *New York Times* predicting "a new era" in the treatment of MS, the recommendation was further affirmation that he had been on the right track.

Stephen Hauser was a junior author on the cyclophosphamide paper, and as a result his reputation was left relatively unscathed. But he believes Weiner may be one of the few who would have the inner strength to keep going after all the public criticism. "As other people fall by the wayside and give up, Howard will not," he says. Hauser believes that the work on cyclophosphamide "inaugurated an era of interest in developing therapies" for MS, previously seen as an untreatable disease.

Still, it hasn't been easy. "It was a major scientific strain," Weiner says, "that I've lived with for over a decade. Many people said to me, 'Howard, don't work with it anymore, it's poison, it literally is poison, it's getting you into trouble.' A lot of the scientific back and forth was very painful."

One of the reasons Weiner was immediately attracted to oral tolerance, as an approach to MS, was that it was so safe in comparison with cyclophosphamide. Oral tolerance involved taking in a substance by mouth that would have virtually no side effects. If it worked, it would silence his critics forevermore.

The idea of oral tolerance has been around since ancient days. In the first century, according to legend, the Mediterranean king Mithridates drank the blood of ducks that had been fed a poisonous weed to protect himself against poisoning by his enemies. The story produced a synonym for tolerance, "mithridatism." American Indians were said to make use of the oral tolerance idea. They would eat the liver of deer, who frequently ingested poison oak, so that they could "tolerate" exposure to the toxic plant in the woods without getting a rash.

Even now, it is a commonplace among herbalists that one can build up a natural immunity through ingestion in small doses over time. A potion made from poison oak is available at homeopathic pharmacies to protect against poison oak rash.

Among the scientists who explored oral tolerance in the first half of the twentieth century was Merrill Chase, whose work with Karl Landsteiner helped lay the foundation for T cell immunology. One 1946 Chase study in particular, "Inhibition of Experimental Drug Allergy by Prior Feeding of the Sensitizing Agent," has become a classic in the field. In this oft-cited experiment, Chase gave guinea pigs low doses of a chemical substance that resembled poison ivy through several pathways. He injected some guinea pigs under the skin, he injected others in veins, and he administered the substance to some guinea pigs by mouth. Then he injected the substance in a larger and repeated dose in all the guinea pigs to see if the initial low dose had affected their reaction. Had any of the guinea pigs become "tolerant" of the allergen?

"Rather unexpectedly," Chase wrote, "and in sharp (but not absolute) contrast to all other methods employed, a blocking effect of substantial degree was found to be induced by feeding the chemical."

Chase had come upon a dramatic manifestation of the phenomenon of oral tolerance.

Forty years later, Merrill Chase's work on oral tolerance became the touchstone for the work of a graduate student at New York University who was studying autoimmune disease in mice.

Cathryn Nagler-Anderson is a researcher in Boston now, at Massachusetts General Hospital. But she was working on her graduate thesis in her native New York City when she made a discovery that began what she now sees as "a revolution in the field."

Like many before her, Cathy Nagler-Anderson was searching for a treatment for arthritis by studying the disease in mice. She studied the males in a particular genetic strain of mice, because they are the most susceptible to the disease. If they are given an injection of collagen, these squirrel-gray mice will often develop arthritis. Usually the disease develops initially in the mouse's hind feet; tiny, pink, almost translucent feet that bear a resemblance to long-fingered human hands. At first, little round balls appear at one or more of the joints. Later one or more feet become highly inflamed and swollen. In the final stage, the destruction of cartilage leaves the digits frozen together and turned inward. The paw is no longer very useful. It is a progression that serves as a credible model for rheumatoid arthritis, a disease that afflicts millions of humans.

Nagler-Anderson was trying to find out if a preliminary small dose of collagen, the disease-inducing substance, could produce "tolerance" in these mice, protecting them from the disease. She tried introducing the preliminary dose by injecting it in various ways, in muscle and in the bloodstream. But she also tried feeding the mice her preliminary dose, using a little ball-tipped needle to slip the liquid into their tiny mouths. That was how she made a surprising discovery: The mice who were fed, unlike the mice who were injected, were then resistant to the disease. There was another surprising finding: If she fed her mice too much collagen, they were not protected. Paradoxically, a lesser dose worked better.

In her paper, which appeared in October 1986 in *Proceedings of the National Academy of Sciences* (PNAS), Nagler-Anderson cited the work of Merrill Chase from four decades earlier. She observed that this was a form of "oral tolerance." But until that time, no one, with the exception of Norman Staines, an English researcher who had published similar findings several months earlier, had thought to apply the principle of oral tolerance to the treatment of an autoimmune disease.

At the time, the discovery was not even the focus of Nagler-Anderson's thesis. "People thought it was kind of funny," she says, "feeding animals antigen. It didn't seem all that exciting compared to what it's become." Now, she says, the "suppression of autoimmune disease by oral antigen has revitalized the field. So many people are working on it now—hardly anybody was working on it then."

But there was one other person who, at about the same time Nagler-Anderson did her work, got the idea of trying a similar experiment in the rodent version of multiple sclerosis, EAE. In 1982, Caroline Whitacre was a young assistant professor teaching at Ohio State University, worrying over how to challenge the students in an advanced graduate seminar in immunology. Her own research for her Ph.D. had been on EAE. As she prepared for her class, Caroline Whitacre got the idea, from an article she read, of trying to prevent EAE in rats through a time-honored method called oral tolerance.

Whitacre wondered what would happen if you fed the animals a small dose of myelin, which contained the protein attacked in EAE, and then induced EAE in those mice. Would the fed protein protect them against subsequent attack? "So we tried a preliminary experiment," Whitacre remembers, "just a very few animals. . . . We didn't know timing, dose, anything." Half of the rats were first fed a milligram of myelin basic protein, then all the rats were given the injection that causes the disease. Rats who are given EAE by injection have a very dramatic response: They become paralyzed within ten to fourteen days, then slowly recover. But to Whitacre's amazement, the animals who had been fed the myelin didn't become paralyzed. Unfortunately, Whitacre was unable to replicate her results right away. "The next ten experiments didn't work. We spent the next two years trying to figure out what had happened."

At the same time that Caroline Whitacre was struggling to replicate her first experiment, Howard Weiner was, as usual, shuttling between his various obligations—the lab, the clinic, scientific meetings. Late in 1985, he had gone to Allentown to work with Larry Levitt on a revision of their book, *Neurology for the House Officer.* On the way home, he was catching up on his reading, as he sat uncomfortably wedged in the narrow seat of a commuter plane, when he came across an article in an immunology journal about oral tolerance. Weiner had actually published a paper on the use of oral tolerance back in 1981, when he was working on reovirus in Bernie Fields's lab. It suddenly occurred to him that there might be a way to use it in treating MS. "For a second, sitting on the plane, I shook my head. Naah! And then five seconds later I said yes but . . . I kept going over in my mind and then it became very clear to me . . . oh my God, this could really be something!"

When he got back to Boston, he spoke to Paul Higgins, a postdoc in his lab who was working with EAE in rats. "I said let's feed myelin basic protein to rats and see what happens." It sounded at the time like a crazy idea. Weiner now tells audiences that "fortunately he was a postdoc, so he had to do what I told him."

In any case, Paul Higgins, a handsome, soft-spoken young biologist, was not the protesting sort. He did as he was asked. There were nine rats in the experiment. Six of them were fed the myelin, then all nine were given the injection that causes EAE. Weiner remembers that he was skiing in Colorado when he got the call from Higgins with the results: The rats who were fed the myelin didn't get EAE from the injection. "They didn't get sick!" Weiner recalls. "I mean, it was very dramatic. And then we said, well, we better repeat this." Weiner had better luck than Caroline Whitacre. "It was positive a second time."

Two years after the first experiment, Weiner and Higgins published a first paper on the phenomenon in the *Journal of Immunology.* The same year, Caroline Whitacre published her first paper, with an almost identical title, in *Cellular Immunology.* Weiner's came out in January 1988 and Whitacre's came out later in the same year, so he had the prior published claim. But no one disputes that Whitacre got there first in the lab.

Howard Weiner often compares research to war, and compares disease to Hitler. Then he explains that to win the war against Hitler, you have to have an overall plan. But then after you have the plan, the day-to-day work, in research as in war, is all about the details. "Some guy is trying to fix a radio," he'll explain, "so that he can communicate with somebody to blow up a bridge." And, as in war, everything in research is a big technical challenge.

The publication of "Suppression of experimental autoimmune encephalomyelitis by oral administration of myelin basic protein and its fragments" in the *Journal of Immunology* in 1988 got Weiner's idea into circulation among his colleagues, but it was greeted with a lot of skepticism. So far he had only described the phenomenon as observed in rats. He needed to explain it. Once again, Weiner was making a controversial claim, one that would take years to prove beyond a doubt.

Weiner was already thinking about how he could try out oral tolerance in humans. "We knew it was going to take years to figure out the mechanism," he explains, "and we didn't want to wait that long to start treating patients." Weiner was also spreading the word and recruiting researchers in other areas to apply it to other autoimmune diseases. If feeding myelin, the protein attacked in MS, could have an effect on that disease, and if feeding collagen calmed mouse arthritis, then there was the possibility that other autoimmune diseases could be treated by feeding the relevant proteins. An ophthalmologist he met in Paris, Robert Nussenblatt, expressed an interest in trying the idea in the treatment of an eye inflammation called uveitis in his lab at the National Institutes of Health in Washington, D.C., and quickly came up with some promising preliminary results in animals. Then one day in May 1988, Weiner met with researchers interested in trying oral tolerance as a treatment for the two most widespread autoimmune diseases: rheumatoid arthritis, where it had already showed promise, and diabetes.

From the moment he got the news from Higgins about the rats, early in 1986, Weiner was sure oral tolerance was worth pursuing. Unlike Cathy Nagler-Anderson and Caroline Whitacre, he was quick to seize on the commercial possibilities. By November 1986, Weiner was talking with a patent lawyer about "the oral feeding," as he called it then. He had exploratory conversations with Dupont and with Athena, a small biotech company that his codirector, Dennis Selkoe, was working with on an Alzheimer's drug. He also began to talk with the administration at Brigham and Women's Hospital about the possibility of partnering with industry.

Nowadays, partnerships between nonprofit research institutions and pharmaceutical companies are commonplace. But they were new and controversial in 1986 when Weiner started to look for backing for oral tolerance. At that time, it was usual for the hospital to collect royalties on patents for technology developed under its auspices. Active participation in the profit sector was considered inappropriate, even unethical. Things began to change in 1980, when Congress passed the Bayh-Dole Act, permitting academic institutions and scientists who worked in them to benefit financially from the commercialization of their discoveries. The Bayh-Dole Act was viewed by critics as a dangerous threat to the integrity of academic research. Al Gore characterized it as "selling the tree of knowledge to Wall Street."

By the mid-eighties, however, some hospital leaders, concerned about the Reagan-era shrinkage of public funding, began to pursue aggressively the private sector partnerships enabled by the law.

One of the most aggressive was Dick Nesson, the innovative president of the Brigham, who organized a group of investors into something called the Venture Advisory Committee, and invited them to come to the hospital on occasion to hear about the bright ideas of hospital researchers. It was at one of those meetings that Walter Channing, a partner in a New York venture capital firm called the CW Group, first encountered Howard Weiner.

CHAPTER 6

The Capitalists

WALTER CHANNING, A SILVER-HAIRED VENTURE CAPITALIST, remembers the day he first encountered Howard Weiner at the Venture Advisory Committee in Boston. "It was a very good afternoon because I met with a neuro-endocrinologist who is one of the world's experts on sleep. And he was extremely lucid, as many of these scientists who teach are." And then, Channing says, it was "next scientist!" Enter Howard Weiner, presenting oral tolerance as a possible treatment for autoimmune disease.

Walter Channing's name and family history made him the ideal intermediary between Wall Street and Brigham and Women's Hospital. The modern Walter was named after his great, great, great-grandfather, an eminent antebellum physician who was himself an innovator. The first Walter Channing was one of the first to use anesthesia in childbirth, and was founder of the first hospital in Boston to provide obstetric care to poor women. Boston Lying-In Hospital, which opened its doors in 1832, was known to generations of Bostonians as "the BLI." The obstetrical services of the Boston Lying-In have long since been integrated into the Brigham, and the 1923 BLI building is now a Harvard research facility. Although neither Channing nor Weiner knew of the connection when they met, the old BLI was about to become Weiner's laboratory, the Center for Neurological Diseases (CND).

Channing found Weiner's presentation "very logical." He liked the fact that Weiner saw oral tolerance as "a way of approaching autoimmune diseases—this was not an MS drug alone but would work in its various modified forms across many diseases." It was, in the language of the business world, a "platform technology," on which researchers might be able to build approaches to a range of autoimmune diseases. What others saw as a problem—the parallels between oral tolerance and homeopathic remedies—Walter Channing saw as a plus. "Whenever a homeopathic phenomenon gets in the hands of real scientists," Channing says, "you know something interesting is going to happen."

Channing is a handsome six-footer with ice-blue eyes and a pursed mouth not unlike William F. Buckley's. He even has a habit, like Buckley's, of patting his gray hair on top with his flattened hand. But, though he looks like the quintessential conservative WASP, he is surrounded, in his wide-open corner office in Manhattan, with evidence of his unconservative ways.

Instead of walls and a door, Channing's space in a New York office building is defined by a cluster of hardwood sculptures he has created. One is of a woman's torso emerging from wood, a hand pressed lightly and suggestively into one breast. Some of Channing's sculpture is playful—he began by carving giant pencils—and some is abstract. But whatever he does, it all grows out of his lifelong love of wood. He fell in love with trees as a boy growing up in exurbia. In high school, he started a tree surgeon business. And after he came to treeless New York, he discovered that they were tearing down a pier near his office and burning all the timbers. Before long, he had a basement on 37th Street full of salvaged wood that became the basis of his art. "A wood person," Channing once said in an interview about his work, "has to be very nonlinear. Ready for anything."

Once, after a hurricane knocked down trees near his place on Long Island, Channing salvaged the huge roots and trunk of an oak tree and had it delivered to his studio on a flatbed truck. After a year of looking at it, he had an "incredible vision" of what it was: "a lion getting a blow-job from a beautiful girl. What was particularly interesting was that the lion was going absolutely *berserk*—Beauty and the Beast, and what he was doing was going to cause him to die from the euphoria of it." Channing made a maquette of his Beauty and the Beast vision. But he was involved in a court battle at the time over custody of his children, so he decided not to risk offending the judge. Instead, he "created the next best thing: an octopus and a sphinx in consort." He notes, a little regretfully, "I was afraid of my own imagination."

The only wall in Channing's sunny office is hung with black-and-white photos of his recent obsession: upended trees. On his 130-acre farm on Long Island, Channing has created a large field of them: slender trunks topped by clusters of gnarled roots, looking like wild and unruly hair. It is an unnerving sight, even in photographs. Next to the photos, Channing has placed a huge variation on his theme: a sculpture perhaps twelve feet high, with tree roots at the bottom and the top. "I don't look at a tree as an object that necessarily has to be right side up," Channing says, explaining the obvious.

Channing was impressed enough with Howard Weiner's presentation to the Venture Advisory Committee to try it out on two of his partners in the CW Group, Barry Weinberg and Chuck Hartman. Hartman, who comes from a chemistry background, was skeptical. "Chemists can usually fault this kind of thing," explains Channing. "They want you to show them the chemistry, which was probably the greatest unknown. What was being observed was a phenomenon in mice." Barry Weinberg, however, agreed with Channing that the idea was worth pursuing.

Walter Channing and Barry Weinberg, the C and W of the CW Group, have been partners since 1968. They are an improbable duo. Channing, scion of a distinguished Boston family, grew up in Dover, a genteel suburb of Boston, and attended Milton Academy and Harvard, followed by Harvard Business School. At the same time, Barry Weinberg was growing up in nearby Roxbury, a then-Jewish neighborhood of modest triple-deckers. After Boston Latin, an exam school, he graduated from MIT in electrical engineering. "I had an uncle who said, 'electronics,'" Weinberg explains. "What did I know?" It turned out he hated engineering. So before long he decided to go to business school, choosing NYU over Harvard because he got a scholarship.

What brought Channing and Weinberg together was disappointment in love. When he was an undergraduate at Harvard, Channing met a fiery and talented Radcliffe undergraduate named Stockard, who was already working on a promising career in the theater. The two married and she became Stockard Channing. But the marriage didn't last for long. When Walter finished his MBA and moved to New York, she stayed behind. The result was that he found himself living in a brownstone on Murray Hill, ideal for a young couple but with "no other half." Barry Weinberg filled the void.

Perhaps because he was working in the computer industry like Channing, Barry Weinberg somehow got invited to Walter Channing's Christmas party. After it was over he announced that he couldn't go home because his marriage was breaking up. So he stayed on the couch for the night. The night turned into a month. Walter remembers that Barry didn't help with the rent but picked up the tab when they went out for Chinese food. Barry remembers that they mostly went out since the refrigerator had nothing in it but beer and "cheese with something green growing on it."

"It was a good time for divorced young guys in New York," Weinberg remembers. Late one night, sipping wine and feeling mellow after an evening on the town, Channing and Weinberg decided they would start a business together. The next morning, Weinberg had second thoughts, but it turned out Channing had already quit his computer job. So Weinberg got $6,000 from a Wall Street acquaintance to finance a study on the computer leasing industry, and Channing-Weinberg was launched. They immediately took $1,000 of the advance and went out to buy a "really fancy hi-fi system." It was before stereo, Weinberg explains. "I remember we bought 'The Doors' and 'Rolling Stones' and played them constantly. We started in his apartment. Then I moved out because I was infringing on his love life."

The businesses have changed, but the team has stuck together. In the early days, they did consulting for financial institutions, helping them make decisions about investing in high technology. Then, as the technology began to move into medicine, they gravitated toward that. They developed a consulting firm for large health care companies, which has since spun off as the Wilkerson Group. Then Channing decided that small companies were going to be very important in health care, and began investing. "He would help them raise money," Weinberg explains, "and he would invest a little of our own money." One of the great technical breakthroughs in health care during those years was in the area of imaging: X rays were being supplemented by ultrasound, nuclear imaging, and CAT scan. When two of the imaging companies Channing-Weinberg had invested in, Elscint and Advanced Technology Laboratories, became successes, they decided to formalize their partnership as a venture capital company.

Venture capitalists are gamblers who count on a lot of failures and a few big wins. Even though the industry has grown enormously since Channing and Weinberg began, its assets are a fraction of those available from commercial banks. Venture capitalists, however, are willing to take risks that banks won't, and to work as partners with entrepreneurs in getting their ideas launched.

As venture capitalists, Channing and Weinberg turned out to have complementary skills. Channing had good instincts about what to invest in, and was good at "starting from scratch and putting a deal together," raising money from people he knew and who knew him. Weinberg was good at managing, and believes that their "hands-on" approach during the start-up phase of new companies has been the secret of CW's success. Weinberg is "very methodical," according to Channing, who is not.

Weinberg's office, just a few yards from his partner's, could be in another building. Channing's unwalled space is bathed in sunlight, and his windows frame a charming row of brownstones along 61st Street. Weinberg's walled office is dark and overcrowded. There are files piled everywhere. His windows overlook the harsh black glass and brass canopy of Trump Plaza—home of a large paint and wallpaper store on Third Avenue. Like Channing, he has his own art on the wall: a strange dreamlike oil painting of a man in an empty room, standing slightly off balance next to a green suitcase with his featureless face floating out toward the viewer.

A small, trim man in a well-made suit, Barry Weinberg stands with his back to the door, staring out on Third Avenue and talking on the phone with someone who wants him to invest in a start-up company.

"Do you have data?" Barry is asking the petitioner. As he listens to the long answer, he turns and riffles the pages of the proposal under discussion, which is sitting on his desk. Clearly, he is not impressed. "We've been in the venture capital business for thirty years," Weinberg tells the caller, "and we invest in the health field only, so we have a lot of experience. . . . How is this different than, better than previous technology?"

Another long answer.

"Your capitalization strategy I think is a little naive," says Weinberg, then adds, without much conviction, "I don't mean to be demeaning."

Weinberg tells the caller to send more data and agrees, reluctantly, to a lunch date in New York.

Once off the phone he admits he has no interest in the proposal whatsoever. "I was doing a favor for a friend." Weinberg has a smile that makes furrows all over his face, and a surprising amount of brown, wavy hair for a man his age.

"We see about 1,500 investment opportunities a year," Weinberg says. "We typically make four or five investments."

Weinberg is in the process of raising $100 million for a sixth ten-year partnership. The money is provided by limited partners, companies, and financial institutions with an interest in the field, and managed by a general partner, in this case the CW Group. Over the past thirty years, the CW Group has been involved, through such partnerships, in starting about seventy companies, and has had to close down three. About thirty have had public stock offerings, and twenty-four others were sold to other, usually larger companies, "not all for profit but mostly." The CW Group is still actively involved with the rest.

When companies have failed, according to Weinberg, it has often been because of the inflexibility of the leading scientist or executive. "We have become very very sensitive to the nature of the person we back," he says. "We've seen many projects over the years in which we were fascinated by the technology and the ideas but did not feel the scientist was suitable for us."

The course of drug development is never predictable, Weinberg explains. "Unforeseen and untoward things happen along the way and you've got to deal with them. And that means not being stiff-necked, not only knowing your stuff but being able to sit down and talk with people and listen." Howard Weiner impressed Weinberg as the kind of person the CW Group could work with—"an honest person, a good scientist. He didn't have a lot of respect for the commercial world, but he observed and was willing to adjust."

All his life Weiner has been a keeper of records. He has videotaped his sons at every stage of their lives and kept a video and audio record of professional events as well. He never goes anywhere without a little notebook or folded square of paper in his pocket, so he can jot down ideas as they come to him. And he keeps journals, both scientific and personal.

Weiner started his first journal at age eighteen.

"Basically," he wrote as an earnest undergraduate, "I'm trying to learn more about myself and this gives an excellent perspective." The journal was his unabashed bid for immortality. "Maybe I'm writing because I want something of me to live on after I'm gone. The only thing I can offer the world is my life, and this is a chronicle of it."

When he resumed his journal twenty years later, Weiner had more to chronicle than his daily comings and goings. Though he may not have been entirely aware of it, the second journal was surely begun with the idea that oral tolerance was going to be the great breakthrough and triumph in his life as a scientist. He wanted to preserve that exciting and gratifying process, for himself and for others who came after him.

But the journal, in the first few years after the oral tolerance finding, more often chronicles frustration and defeat than triumph. "Most things in research are difficult," Weiner is fond of saying, "most things don't work. Otherwise it wouldn't be research." Sometimes, in those years, it seemed that everything in research was difficult.

The first step was putting together a team that shared his hunch about oral tolerance. Even though the work was "very exciting," he wrote in his journal, "Paul Higgins doesn't realize it." Within a year of the first breakthrough, Higgins was talking to Weiner about the possibility of leaving to work in a biotech company. He found the oral feeding repetitive, he told Weiner, and wanted to do other things. Weiner, aware that Higgins was "all I have on feeding," resolved to "get tough on him and be explicit that I want him to move up a level." But in the end, Higgins informed Weiner that he wanted to leave, and Weiner concluded that it was for the best.

"In a way I am happy," Weiner wrote in his journal, "as it will force me to take over more control and monitor the scientific experiments carefully." Weiner was aware of the need to deepen his thinking. "How *does* oral tolerance work?" he asked himself. Uncharacteristically, Weiner admitted to feeling "a bit inadequate" to all he was faced with.

The oral tolerance idea thrust Weiner into a whole new area of immunology: the immune system of the gut. Nowadays, his talks often begin with a description of this unique system, which he describes as "a tube running from our mouth to our rectum." It is, he explains, "the largest immune system in the body," and the one that, more than others, is set up to accept foreign antigen rather than react against it. It is, by nature, a generator of tolerance. But when he published his first paper on oral tolerance, Weiner knew little about the mucosal system of the gut, which has since become the focus of much of his lab's research.

Nor was oral tolerance a fashionable idea in the scientific atmosphere of the eighties. "We were ridiculed," Weiner says now, looking back. It was a low-tech approach in a high-tech era. Scientists criticized it because there was no explanation of how oral tolerance might work on the molecular level. And "molecular" was the "in" word at the time. The focus in biotech was on designing drugs, not triggering natural physiological processes. What Weiner seemed to have done was to find a shortcut: a way to stimulate the animal's own T cells to suppress autoimmune disease. But he badly needed explanations to back up what seemed to many in the field to be preposterous claims.

The beginning of a change for the better came with the arrival of Ofer Lider, a scientist Weiner had first met at the Weizman Institute on a visit to Israel in 1986. Because his wife Mira was Israeli, Weiner had developed a strong connection to Israel over the years. He did a one-year internship there after medical school, and returned often with Mira to visit her family. His Hebrew was fluent. So it was natural for him to look to Israel for collaborators—and Ofer Lider was the first of several researchers who would travel to Boston to work on oral tolerance. "I will love working with him," Weiner predicted even before his arrival.

Very quickly, Lider and Weiner established a rapport. Lider reminded Weiner of his younger self. Dinner at the Liders included two little girls, aged five and eight, running around in their pajamas, like his sons in an earlier stage of life. Weiner took Lider to see a Boston Celtics game that was won, Celtics-fashion, in double-overtime. Very quickly, too, there were exciting results in the lab. As Lider carried on with Higgins's work, he discovered that the animals, once fed the protein to induce tolerance, produced a cell that could be isolated and seemed to be a "suppressor cell" that prevented them from getting sick with EAE. In October 1987 Weiner wrote in his journal that Lider "clearly has suppressor cells in vivo." Then he adds: "Could be major."

The frustration for Weiner, at this point, was that his lab had already embarked on other projects that had to be seen through to completion. "When you make a major breakthrough, as we have," he observed, "you have to keep focused on it. A lot of people have made that mistake. You need to be sure you understand what you've discovered." Faced with a temptation that wasn't relevant, Weiner would remind himself, "you can only dance at so many weddings"—a variation on the wise Yiddish saying, "you can't dance at two weddings with one tukhes."

At the time of the oral tolerance finding, Weiner was dancing at quite a few weddings in his search for a cure for MS: He was finishing up work on plasma exchange and had experiments going on with cyclophosphamide. He was collaborating with Irun Cohen, another Israeli researcher and friend, on a possible vaccine for MS. There was even a researcher in his lab working on polio virus—a last link to his work on viruses in Bernie Fields's lab. He was also struggling with the three plagues of a lab director's existence: space problems, money problems, and problems with personnel. "There is too much pain in my life," Weiner complained to his diary on one occasion. And on another, "I do not enjoy my days presently and nothing is worth that."

The stress made Weiner worry about his own health. He even developed a fear that he himself might be developing MS. Running around the reservoir near his house one day, he noticed some weakness in his legs. A blood test added to his fears: There were some changes that paralleled the changes his lab was observing in MS patients. "Should I get an MRI?" he wondered. It is not unusual, Weiner notes, for physicians to have such concerns. "You see all these people with this illness and everybody says to you, 'I never thought it would happen to me.' So you know you're not immune. You know that there, but for the grace of God, go I." Weiner, however, turned out to be among what psychiatrists call the "worried well."

During this time Weiner had seemingly endless meetings with the Brigham hospital administration over finding more lab space. He was particularly annoyed with the administrator who encouraged him to "think programmatically" and then "wound up counting square feet." For a while, it looked as though he'd be able to move the entire lab to another building. When that fell through, the disappointment was "almost unbearable."

Meanwhile, one of Weiner's postdoctoral fellows submitted a paper to a journal without telling him—grounds for dismissal in many labs. When confronted, the postdoc told Weiner that he did it because Weiner sat on things too long. Another postdoc didn't "write papers or do his own work," and expected to be supported indefinitely anyway. Weiner hired an administrator who was supposed to take charge, but instead offended the secretaries, who threatened to quit. One did quit, leaving a hostile note pointing out to Weiner that his grants were "overspent."

That secretary knew just how to get to Howard Weiner. The biggest preoccupation for him, as for every lab director, was the constant worry about money. The cycle of grant-writing and grant anxiety was as inevitable as the seasons. There would be a period of very late nights and weekends when Weiner worked with others in the lab on grant applications to the National Institutes of Health (NIH) and the MS Society. Then there would be what he described as the "nice limbo" after submission: "work accomplished, no pain of rejection or disappointment." Then came the agony or ecstasy of the response.

April 1987 proved to be a particularly cruel month for Howard Weiner: two grants he was counting on from the MS Society were rejected, and one of them was given an "atrocious" review that he believed was unfair and vindictive. "My confidence, energy, and dedication is shaken," he wrote in his journal. "I don't know how much longer I can handle these things. The work itself and taking care of patients is taxing enough. I only hope the cumulative effect will not be causing a mortal blow to my psyche and energy."

Meanwhile, Weiner was expending quite a lot of energy courting donations from wealthy businesspeople. He traveled to Texas to visit with an oil millionaire who had MS in his family. And he spent a great deal of time staging a presentation in New York for Milton Petrie, a billionaire who had been introduced to him by a well-connected MS patient. Petrie was considering a major gift to Harvard hospitals that could greatly benefit Weiner's lab. In early October 1987, Weiner and Dennis Selkoe convinced Dick Nesson, president of the Brigham, and Eugene Braunwald, chair of the Department of Medicine, to travel to New York for a presentation to Petrie. But ten days later, the stock market crashed and Petrie changed his mind.

Weiner vowed around that time that he was going to quit trying to raise private money and get back to doing things for "the love of science." Yet how could he? There were massive bills for purchasing and maintaining experimental animals that had to be paid. There were nagging worries about space. And there were more general money problems that needed fixing.

Money, of course, was just a means to an end. For Weiner, the ultimate measure of success was not how much he could raise but what kind of results he got in his lab. What he cared about, more than money, was reputation. One of the cornerstones of reputation, for him and for his lab, was being named a full professor at Harvard. This, too, was a sore subject during this period. Even before his name was first submitted for a professorship, Weiner had been chosen, after a national search, to receive an endowed chair at Harvard for multiple sclerosis research. That chair was given by the Kroc Foundation, established by the family that built the McDonald's hamburger chain. The professorship, however, could be conferred only by a committee of his Harvard peers. And Harvard, as one colleague observes, "appoints people to a professorship when they can't find a reason not to." Howard Weiner put together his first application in March 1986, when he was forty-two, and received his first rejection the following May. "When compared to other 'neuroimmunologists' in the world," he noted in his journal then, "I didn't rank high enough on all the lists." Weiner told himself it "makes no difference" since he had the Kroc chair anyway. But at the end of that day's diary entry he added, in parentheses, "It hurt nonetheless."

Given all the frustrations Weiner was encountering in the nonprofit sector, it's not surprising that the possibility of a partnership with capitalists was seductive. Peers who reviewed his grants looked for flaws that could allow them to reject them. Others who considered Weiner for a professorship looked for some reason to turn him down. And hospital administrators doled out space in square feet. Channing and Weinburg, on the other hand, looked for possibilities and overlooked technicalities, and when they talked about money, they talked in millions. One day in August 1987 provided a particularly stunning contrast between the two worlds. On August 16, Weiner learned that a grant application for polio research had been turned down by the NIH because "they didn't think I knew molecular biology." The rejection put him in such a foul mood that even the parking lot attendant noticed. "It couldn't be that bad," the attendant told him as he arrived at the lab the morning after. What made it all the more unbearable was his realization that the rejection was the result of his own inattention to the application. He told himself, once again, that he must "only do things that I can do 100 percent."

That day, as he castigated himself about the loss of a grant for $150,000 a year for three years, the CW Group offered him $2.5 million, to be spent pursuing oral tolerance for the treatment of autoimmune diseases. For the first time ever, Howard Weiner glimpsed the possibility that he might not have to worry about research money all the time.

But nothing comes without strings. Very soon, Weiner was faced with a conflict between the ethic of openness, which rules in academia, and the proprietary mentality of the profit sector. In December 1987, Weiner traveled to his hometown of Denver to present his work to the first meeting of the Scientific Advisory Board of what he then called "The Brigham Company." It was a small group: Weinberg, several representatives of the hospital, and one veteran of the pharmaceutical business. But the response was very positive. And even though the deal with the CW Group was not yet final, everything looked promising.

There was worry, however, about publication of the work in a scientific journal. Two years had passed since Higgins first observed oral tolerance in rats. But, because of the usual lag time between discovery and publication, the first paper on the discovery was about to come out in the *Journal of Immunology*. In fact, Weiner had the galleys of the journal article in hand when he went to the Denver meeting. "I am confronted with a major issue of publication of Paul Higgins' paper," Weiner wrote in his journal. "Should it be published? What if it blows the deal?"

A scientist's reputation rests on having his research taken up and duplicated by other laboratories. For that reason, the researcher tries, in his published work, to be as precise and complete as possible. A biotech company, on the other hand, needs to have its "technology" (as research is usually called in the business world) protected. Since it takes ten to twenty years, and from $100 million to $300 million to develop a drug, the only way a biotech company can recover its costs is by having patents, which protect the technology for twenty years from the date of filing. If the idea pans out and turns into a drug, the biotech company needs to make enough money during the years remaining on the patent to offset the costs of development. Once the patent expires, generic drugs will cut sharply into the profits. Critics argue that the pharmaceutical industry inflates claims about the cost of drug development, in order to charge exorbitant prices for drugs under patent protection. But the patent system, for better or worse, is the sine qua non of drug development in the profit sector.

Even before Channing and Weinberg became interested in oral tolerance, Weiner had consulted with the Brigham's lawyers in Washington, D.C., and taken out a patent, assigning the invention to the hospital. This had long been standard procedure at the Brigham's technology transfer department. The CW Group, however, wanted to consult their own patent lawyer, Peter Ludwig of the venerable New York firm of Darby and Darby.

"Barry Weinberg got involved and wanted to get an independent view," Ludwig explains. "And he came to us. Barry asked us for our view on how strong we thought the existing patents were. Were there any references, articles, old books, or other patents that might have made what Howard Weiner had done unpatentable? Or require the protection to be narrowed down? That's a pretty standard look-see for us."

Peter Ludwig's smile is a pleasant surprise, friendly and a little impish because of the space between his front teeth. When he speaks, his deep voice resonates like the tones of a string bass. "We've worked with a lot of venture capitalists, so we've developed a little niche in what's called patent strategy. It's not so much, 'can you get a patent?' but 'how can you take the patents you have and sort of maneuver them around to make them last longer, to work out the maximum protection in the United States and Europe?' There are all sorts of schemes."

Ludwig's large, sunny office, on the 27th floor of an east side Manhattan office building, has a window that looks out over the East River to Queens and beyond. On the wall opposite, a photo collage illustrates how Darby and Darby established its reputation. It shows Samuel Darby with the electrical engineer Lee DeForest. DeForest, who has sometimes been called "Mr. Radio," invented the triode electron tube, making it possible to detect and amplify radio waves. The inscription, by DeForest to Darby, reads "To my best friend and pal in countless patent battles."

Appropriately enough, Peter Ludwig was himself a ham radio operator when he decided to pursue law, partly because he needed a haven from the Vietnam War. He was only two years out of law school when he got his first academic clients—two researchers at Columbia who wanted to start a little pharmaceutical company. Since then, he's learned a thing or two about working with start-ups.

"The problem in most small companies, believe it or not, is not having too many patent applications. It's having too few." Especially in biotech, where the workers are highly educated, there is a tendency to think that there's nothing new or remarkable about what they're doing. "They'll say, 'You know, anybody could have done this, you had to work a little bit but it's pretty obvious.' It takes a long time to persuade these folks that although it may seem simple to them, it's a big deal." Ludwig prefers to get "a flood of patent submissions. Then it's easy to decide which are noteworthy and which are peripheral. The challenge is getting the spigot open and the water flowing."

The crucial measure of patentability, Peter Ludwig explains, is whether the idea is obvious to a person of ordinary skill in the field of the invention, at the time the invention is made. In the case of oral tolerance, a naturally occurring phenomenon, there must be evidence of the "hand of man" manipulating what occurs in nature. "If something is truly a product of nature we need to do something to it, isolate it, purify it, to make it patentable. Picking an apple off a tree is not a patentable idea, since it's an obvious thing to do. But how about isolating the gene that encodes insulin in your body? Can you pick one of those off a tree? The way of isolating the gene is the invention."

Ludwig has very little patience for the ethical issues raised, particularly in Europe, about patenting products that are derived from the natural world. In the United States, it has been relatively easy to procure patents for genetically engineered mice, for instance. The oncomouse, altered so that it gets cancer, was easily patentable in the United States, but has been the subject of a major dispute in the European patent office about whether a patent claim was moral. "What is morality doing in the patent law?" he asks. "It's the wrong damn place for it." Yet even biologists who support patenting useful manipulations of natural phenomena question the idea that private corporations could lay claim to parts of the human genome, the genetic legacy of the human species. Now that the map of the human genome is complete, even U.S. patent law may well be forced to deal with questions of morality.

As for the conflict between the research model of full disclosure and the business model, the key is, once again, patent protection. "There is a certain tension there," Ludwig says, "but the way it works out is usually pretty simple. The researcher agrees to delay publication of the paper for thirty or sixty days until we can file a patent application. Then they're free to publish away."

Problems arise, Ludwig says, because small companies, accustomed to the university environment, have "a tendency to publish indiscriminately." Ludwig illustrates with a dialogue. On Friday afternoon, he gets a phone call from a scientist: "I'm off to the American Diabetes Association and I'll be delivering a paper and I just thought you ought to have a copy."

"When is that?" Ludwig asks.

"Oh tomorrow, tomorrow morning," the researcher answers.

The publication of an abstract of the paper has undoubtedly appeared already in the program for the diabetes meeting. The new ideas in the paper may be safe in the United States, since U.S. patent rights are awarded to the inventor, the first person to come up with the idea, rather than the first person to file a patent application. But patent rights in the rest of the world are based on who files first. "That little abstract blows off all of the European and Japanese patent rights unless you run and file a patent."

In such a case, Ludwig says, there is only one solution: "work all night."

Ludwig makes the tension between academic openness and secrecy in the profit sector sound minor. But that isn't always the case. A survey of Boston researchers conducted by Dr. David Blumenthal suggests that one in every five medical scientists in the United States has delayed publication of results for six months or more to protect financial interests. The study was accompanied by a case example, "a cautionary tale that illustrates the sharply differing views of research . . . by a university researcher and the company sponsoring that research, if the company's product is at stake." Knoll Pharmaceutical Company claimed that its drug for treating thyroid disease, called Synthroid, was superior to the three other drugs on the market. But a study conducted by University of California researchers showed no demonstrable differences between the drugs. Knoll engaged in a long campaign of legal threats and intimidation of the California researchers to block publication of their results.

Fortunately, in the case of Howard Weiner's article for the *Journal of Immunology,* Ludwig decreed that patents had already provided enough protection. On December 14, Weiner noted in his journal, after a meeting with Walter Channing, that "the article doesn't appear to be a major issue." He sent the corrected galleys off to the *Journal of Immunology.*

With his usual optimism, Weiner wrote in his journal around the time of publication that "a new company is about to be born." But the birth was actually nine months away. Before they were willing to sign on, the CW Group had some more questions they wanted answered. Their questions were practical and global: "Will this work?" And more particularly, "Will it work as a drug for humans?" And finally, "How long is it going to take before we can start testing in humans?"

Here again, corporate and academic research cultures sometimes clash. The scientist wants to answer the basic science questions. The investor wants to leapfrog basic questions and get to human trials. Nobel Prize–winning biochemist Arthur Kornberg has argued that the best way to get results is to put practical goals aside. "It may seem, even to many scientists, unreasonable and impractical (call it *counterintuitive*) to address an urgent problem, such as a life-threatening disease, by pursuing apparently unrelated questions in basic biology or chemistry. Yet the pursuit of curiosity about the basic facts of nature has proven, throughout most of the history of medical science, to be the most practical and most *cost-effective* route to successful drugs and devices."

The issue of basic versus goal-oriented, "targeted" research is debated endlessly in scientific circles. Dennis Selkoe, Howard Weiner's partner at the CND, suggests that "you need both," and that the dichotomy between the two is false. Even Arthur Kornberg, although advocating basic, untargeted research, is the founding scientist of a successful biotech venture called DNAX that has "milestones," dates by which certain research goals will be reached, just like every other biotech company. Kornberg insists that "milestones are a charade," but his company continues to set them anyway.

Howard Weiner, who has sometimes been criticized by his colleagues for being too eager to get to results, had a natural sympathy for the goal-oriented thinking of the CW Group. But even he was surprised, in the early days of negotiation, at just how serious the venture capitalists were about setting "milestones." In one talk to an interested group of investors, Weiner said it would be possible to advance to clinical trials in human subjects in twenty-four months. In the next talk he said it would be thirty months. Later he learned that Walter Channing was quite alarmed by the six-month discrepancy. This was the first indication of a lesson Howard Weiner was going to learn and learn well: When investors have money in a project, the meter is ticking.

The delays in finalizing the deal with the CW Group were frustrating to Weiner. By May, he wrote in his journal that he was "beginning to get tired of the entire process. Another two weeks, another two weeks, and always something needs to be done 'immediately.'" Besides the CW Group, there were other potential investors from Morgan Stanley and New Enterprise Associates, who came poking around the lab and calling up with often naive questions.

Nonetheless, even before the deal was final, Weiner's spirits began to lift. The Neurologic Institute of the NIH awarded him a prestigious Javits grant. Named after Senator Jacob Javits, who died of a neurologic disease, the grant is given to leaders in the field. It provides them with about $150,000 a year for seven years. In March, Weiner wrote in his journal: "Most things are in place . . . life can truly be enjoyable now. With Javits grant and Company and money raised, I'm sitting pretty." A few months later, with the CW Group deal still pending, Weiner presented the oral tolerance idea at a scientific conference for the first time. Until then, oral tolerance had been one of a number of research avenues he was pursuing. But he concluded, after the presentation, that oral tolerance should be his main focus from that time on. "The oral tolerance could be my last hurrah," he wrote. "It will require five to ten years of work."

What finally clinched the deal with the CW Group was the success of an experiment the venture capitalists had requested. When Channing and Weinberg asked their scientific advisers about Weiner's idea of oral tolerance, the advisers' reservations concerned the mouse model of MS, EAE. Did the positive results in mice predict success in humans? Channing and Weinberg reasoned that positive results in two animal diseases would be better than positive results in one. So they asked Weiner to test his theory by feeding a protein to a strain of mice, called NOD mice, who naturally contract another autoimmune disease, diabetes. It took a lot of trial and error to discover which protein to use and in what quantity to feed it. Finally, on August 12, 1988, Weiner wrote in his journal: "An amazing day in the laboratory in terms of results: It looks like we have prevented diabetes in the NOD mouse by feeding pancreas. Results are reminiscent of initial EAE results. . . . This could be a major breakthrough."

Walter Channing remembers getting the call from Weiner about the results. "I was in Bridgehampton sculpting, as I often am—I have more dirty clothes than suits. And Howard was very excited, because he was getting all the right results on the diabetic mice. The data was starting to look pretty good."

In September 1988, following the positive results in the diabetic mice, Weiner and the Brigham finally closed the deal with the CW Group. Around the same time, the name of the new enterprise was changed from "The Brigham Company" to "Autoimmune Technologies." Howard Weiner wrote in his journal, "Autoimmune Technologies is off and running!"

CHAPTER 7

Off and Running

Cow Brains

One of the pleasures of the oral tolerance idea, for Howard Weiner, was that it forced him, in middle age, to delve into a whole new area of immunology. He liked studying for an afternoon at the Countway, Harvard's medical library, surrounded by medical students and residents. He enjoyed going to immunology conferences where the material was unfamiliar, instead of neurology meetings, where he often felt dissatisfied. He spent an "exhilarating week" at an immunology conference in Berlin, attending symposiums and workshops nonstop. By the end, he felt he had "moved my understanding of immunology up one entire level." On the way back on the plane, he was tearful, thinking that the oral tolerance idea "is going to work and we will have a treatment for MS."

But as he began to put together plans for the first human trial of oral tolerance, the unknowns were daunting. Even before the deal was finalized with the CW Group, the Scientific Advisory Board of the company-to-be struggled with how to go about it. How many MS patients should be in the first trial? Should they be in early stages of the disease or late? Even more rudimentary, which of the myelin proteins involved in the MS inflammation should be fed to humans in the first trial? And where should the myelin come from? From pigs? From cows? There was also the possibility of making a human version of myelin, using recombinant techniques. But this would be more costly and time-consuming than using myelin from nonhuman sources. David Hafler, who had particular expertise in human immune responses to myelin proteins, joined the board in that early stage, and added his scientific expertise to the deliberations.

In the end, it was decided to feed myelin proteins extracted from the brains of cows. But the proteins were not something you could purchase: The lab had to figure out a way to make it. So at the same time that he was learning about immunology, Weiner was also learning, along with a worker in his lab named Ahmad Al-Sabbagh, about where to find cow brains.

"We had a slaughterhouse in Lynn," Ahmad Al-Sabbagh remembers, "and we used to call on this guy—John LaRusso was his name—and get an order of fifteen, twenty, twenty-five brains. He'd bring them in all bloody and everybody looked at them disgusted. And we would sit there and clean them. We did it in the lab because we had no other space to do it in. Then you package them and make sure they're tightly frozen, because the protein can degrade and lose its activity."

Ahmad Al-Sabbagh grew up in Lebanon and earned his medical degree in Spain. His wife, whom he met there when she was a graduate student, comes from Haiti. His children go to Catholic schools and observe Ramadan. But even he, despite his ecumenicism, was a little nervous about the first challenge he faced when he came to the Weiner lab. He was hired to work as a technical assistant to Ofer Lider, an Israeli. Ahmad Al-Sabbagh had never met or spoken with an Israeli when he came to the CND. "Coming from Lebanon, it's just the way you grow up," he explained.

At their first encounter, Ahmad hoped that "he would look at me, at my credentials, not where I'm from." Lider turned out to be warm and welcoming. He even knew a few words of Arabic.

And what's more, says Ahmad with a twinkle, "he looked like me!"

The Lebanese and the Israeli worked well together. But Ahmad's deepest allegiance was to Howard Weiner. His large, dark eyes well with feeling when he talks about it. "Howard has always been my mentor," Ahmad says. "I consider him—I don't want to say as a father because he's not that old—but he has a father-type relationship to me. He is a very dear person to me."

Al-Sabbagh got close to Weiner when he took on the messy job of extracting myelin basic protein (MBP) from animal brain. "I worked with Howard from the beginning," Al-Sabbagh says. It was uncharted territory in the lab. At first, researchers in the lab needed myelin for their own animal experiments. "Howard even offered a dollar for each milligram of MBP anybody could get," Ahmad remembers. "The first time I made it in a small quantity, I ran into Howard's office to tell him. He said, 'make it again.'" Soon Ahmad was making large batches of MBP and other myelin proteins from the brains of several different animals. Other labs had established the protocol for making MBP, but no one had been able to produce it in such quantity.

When it came time to prepare myelin for human consumption, Ahmad was in charge. But because it was so crucial and so new, Weiner took an active part in the process. In August 1988, Weiner wrote in his journal: "spend time helping Ahmad making gels and preparing myelin . . . I need to be part of it." And a few weeks later, Weiner, Al-Sabbagh, and Hafler went into Boston's Chinatown to visit a small company called BioPure, which had the expertise and equipment necessary to extract myelin from cow brains for the human trial.

Five months later, in January 1989, Weiner, Al-Sabbagh, and Hafler loaded up Weiner's car with frozen cow brains and large containers of sucrose to be used in the processing, drove to Chinatown, then transferred the containers to a small elevator that carried them up to the modest third-floor BioPure laboratory. There a technician named Bin Wong began the process of extracting the myelin to put in capsules that would be given to patients. While Weiner and Al-Sabbagh were there, things didn't go so well: The machine that was supposed to do the job balked. At midday, they took a break for lunch in Chinatown.

Two weeks later, though, the myelin arrived at the lab, ready to be packaged in pills, and fed in large quantities to guinea pigs to test for an adverse reaction. And a week after that Weiner brought capsules to the meeting of the AutoImmune Technologies Scientific Advisory Board. There, with the pills before them, the board made the final decision on the shape of the first trial of oral tolerance in humans. It would be a year-long trial of thirty patients in the relapsing and remitting stages of multiple sclerosis: Half the patients would be given the myelin from cow brain, and half would be given a placebo made with lactose. Howard Weiner, because of his close association with the oral tolerance idea, would not be one of the Brigham doctors conducting the trial. It would be a double-blind trial: neither the patients nor the two doctors administering the pills would know which pills contained active drug and which contained an inactive substance.

By fall the pill, a red capsule that was either an inactive placebo or a dose of myelin, was ready to be tried out on patients. Howard Weiner went before the Institutional Review Board (IRB), a committee of physicians, health workers, and laymen who must look over and approve every clinical trial conducted at the Brigham, to present the plan. The IRB gave him permission to proceed. On October 3, 1989, the first pill was given to a patient in the trial. "It could be the beginning of something monumental," Weiner wrote in his journal. And then added a worry: "I just hope people don't get sensitized and have more attacks."

The Last One In

Rick L., a slight computer programmer, was newly married and only twenty-nine years old when he began to have problems with his vision. Later there were other odd symptoms, which seemed to come and go: Sometimes he had trouble walking, sometimes he couldn't raise his hands above his head to shampoo his hair.

His doctor suspected multiple sclerosis but wanted another sign before he made the final diagnosis. Then, on Labor Day 1989, Rick's eyes did something they'd never done before. "One of my eyes," he remembers, "looked straight ahead and the other eye was looking off in the distance somewhere." The doctor told him that was a classic symptom of MS. "That closed the book."

Rick and his wife, who lived in Hartford, Connecticut, at the time, sought a second opinion from a specialist at the University of Connecticut, who confirmed the diagnosis but had no new treatments to offer. He mentioned, though, that there was a group in Boston trying something, and suggested that Rick get in touch with Howard Weiner.

Rick L. will always remember Halloween that year. It was the day he and his wife closed on their new house—a memorable day in the life of any young couple. But for Rick and his wife, who had recently had their first child, the excitement was laced with terror. That morning, Rick couldn't manage to tie a necktie. It wasn't a problem of manual dexterity but of memory, which is sometimes affected in MS. He stood in front of the mirror trying to make a Windsor knot but he couldn't remember how to do it. Instead of a tie, he wore a crewneck sweater to the closing.

In November, not long after he and his wife bought the new house, Rick went to Brigham and Women's Hospital to see David Dawson, one of the neurologists conducting the oral tolerance study that Weiner initiated. "Dr. Dawson looked me over and said, 'Yeah, you'd be a prime candidate for a study we're just starting.' I became the thirtieth of thirty people, the last one in. I started in February of 1990."

Cow Noses

Not long after the trip to BioPure with the cow brains, Ahmad Al-Sabbagh paid a visit to a slaughterhouse with two other scientists,

David and Rosie Trentham. Instead of brains, they were looking for cartilage from the noses of calves, because it is rich in collagen, the protein that was going to be fed in an oral tolerance trial for rheumatoid arthritis. "Each one of us had an axe," Ahmad remembers, "and we went into the freezer, a refrigerator bigger than a room. And we all had to put on coats and masks and gloves." The goal was to extract the cartilage from the head in one piece. "You became a butcher. You had to be careful because all the pieces were flying all over the place. And Rosie was saying, 'Be careful with the cartilage!'"

The arthritis of aging, called osteoarthritis, can be painful and inconvenient. But the autoimmune form of the disease, rheumatoid arthritis, is more debilitating. It occurs when the body's immune system attacks synovial tissues in the joints, resulting in a progressive, painful inflammation, along with crippling deformation of the hands, feet, hips, knees, and shoulders. The many drugs available for treating rheumatoid arthritis have side effects and mixed results. Also, arthritis had been the first disease to respond, in the animal research of Cathy Nagler-Anderson and Norman Staines, to the oral tolerance approach. So it was natural for Howard Weiner to look, early on, for a collaborator in the field of arthritis who could start to do what he was doing with multiple sclerosis. David Trentham, a researcher at the neighboring Beth Israel Hospital, was already doing work in rheumatoid arthritis with his wife, Rosie, and seemed a natural choice.

Weiner first met with Trentham in the spring of 1987 and reported in his journal that Trentham was "quite excited" about the idea of working together. "He is ready to feed collagen to patients in the fall. He is perhaps a bit too anxious but I think he will be a good collaborator. Can't do it alone."

It was Trentham's idea to choose collagen, which is present in the joints, as the protein to feed in a human trial. "There were several candidate antigens on the table," he remembers. "My prejudice and my long-term research niche has dealt with the probability that, at least in some patients, [rheumatoid arthritis] is propogated by an aberrant T cell response to cartilage, or type II collagen." If collagen is a cause of the inflammation, then the logic of oral tolerance suggests that it may be possible to feed it and get suppression. That is why the Trenthams, along with Ahmad Al-Sabbagh, went looking for cartilage in a slaughterhouse.

The Unblinding

The weeks leading up to the unblinding of the first human trial of oral tolerance for MS were extremely tense for Howard Weiner. The United States was at war with Iraq, and Iraq was retaliating by firing SCUD missiles into Israel. The Weiners spent hours watching events on television and calling Mira's family in Israel. In addition, Weiner agreed, at the last minute, to fill in on the neurology ward at the hospital, after a colleague drowned tragically in Miami. He also had grant deadlines that kept him up one night until four in the morning. And it was hard for him to stop worrying about the trial results. "Could be a major blow to the program with negative results," he noted in his journal, "hard to maintain momentum."

On March 11, 1991, a little over a year after the last patient, Rick L., entered the study, Weiner convened a meeting in his office to hear the statistician's report on the results. The two doctors who had conducted the study, Dave Dawson and Glenn Machin, were there, wondering if their various guesses would turn out to be right. There were, in particular, two male patients in the group of thirty who had been on a downward course before the trial, but had done extremely well after they started taking their trial pills. Dawson and Machin told Howard Weiner that if these two patients were on active drug, then maybe there was something to oral tolerance. If they were on placebo, however, he should probably forget the whole idea.

The statistician told the group assembled in Howard Weiner's office that the news was good. Although the sample was too small to provide statistically significant results, there was no doubt that a number of the patients did well on oral myelin. Of the fifteen patients on oral myelin, only six had a significant attack. Among those on placebo, twelve of fifteen had an attack. The two men whose disease had been arrested were both on myelin, and men in general did better than women. David Hafler later reported that there was a decrease in MBP-reactive cells in the patients fed the oral myelin, evidence of a biologic effect.

Howard Weiner was euphoric. In his journal he wrote a long rumination under the heading in capital letters: POSITIVE RESULTS IN MYELIN TRIAL! My life is now changed," he wrote. "May have actual breakthrough in MS. . . . Of course, there is a long way to go, but there is now a real chance. I think I will now work on this for the rest of my life. A burden has been lifted . . . so many patients have been counting on me to come up with something. . . . I was afraid I would let them down. . . . Judgment from the world of biologic realities . . . not from peers, committees, back-biting scientists, or supporters who will always compliment you. Let nature be my judge." All the apprehension he felt leading up to the unblinding was gone. "Oh Lord, I feel so wonderful right now," he wrote.

When Rick L. got the news, he was excited too. Right after he started taking the pills, he had an attack that made him fear that the medicine wasn't working or that he was on placebo. But then things began to settle down. During the previous nine months, the attacks had stopped almost entirely. When the study was unblinded, he learned that he had been on active drug. Naturally he elected to keep taking it.

The First Employee

March can be wintry in Boston, but this particular March Thursday in 1992 was mercifully mild. After Howard Weiner and Robert C. Bishop talked over lunch at Davio's, a sleek Italian restaurant, they walked the mile or so back to Howard's lab. Bishop was the latest in a line of candidates for the job of CEO of AutoImmune, Inc., as the company was called by then. He was a Californian, unused to frigid weather. But as he breathed in the balmy air, he asked himself, "What's wrong with this? Why wouldn't you want to live in Boston if it's like this?"

Even before the walk, Bob Bishop was feeling positive about the possibility of becoming CEO of AutoImmune, Inc. He'd read the relevant scientific papers before he arrived at the lab that morning, and had a lot of questions for Howard Weiner. During the meeting in Weiner's office before lunch, he'd been impressed with Weiner's answers. "He wasn't evasive. When he didn't know he said he didn't know. He wasn't overly trying to sell me. I think he was trying to understand how much I knew at the same time." Most important, the chemistry felt good. "Within five or ten minutes there was a perceptible click. You just knew this was somebody you could work with."

But it was going to be hard for Bishop to leave California—and not just because of the weather or his current job. Few in his family have strayed far from California. Bishop's parents met at the University of Southern California, and he met his wife there too. Bishop's four siblings still live in California, and all the generations get together for large family gatherings every couple of months. For Bob Bishop, leaving his job was easy compared with leaving the clan. And it was going to be tough for his wife Susie and for Brian, the one of their three children still at home.

Before he met Howard Weiner, Bob Bishop had had an interview in an airport hotel with Barry Weinberg, who had been playing the role of CEO ever since AutoImmune became a company over three years earlier. Weinberg had been drawing on his circle of investors to finance research in the Weiner lab, as well as preliminary research in David Trentham's lab on rheumatoid arthritis and at the Joslin Center on diabetes. He had also managed to initiate negotiations with the pharmaceutical giant Schering-Plough, which was considering support for the arthritis research. By the time Bishop came along, Weinberg had investors from his circle who had committed $8 million to keep the research going. But Weinberg's strategy, with AutoImmune as well as other companies the CW Group backed, was to turn the company into a viable entity so that its shares could be sold to the public. To launch an initial public offering, or IPO, he needed to have a CEO who could take charge, formulate a strategy, and hire a team of officers.

Weinberg was tired of his stand-in role as chief AutoImmune executive in any case. It was taking up a lot of time, and required getting up early for frequent commutes to Boston. He'd engaged headhunters, including one who promised he could get "an audience with the pope." But until Bishop, all the candidates had been problematic. There had been an interview in 1989, over lunch at the Bay Tower Room, with a candidate who was then the CEO of another company working on immune disorders. But he was pledged to protect secrets from that company that could tie his hands at AutoImmune. Another candidate seemed ideal, but insisted on moving the company to California. Weinberg wanted to comply, but Howard Weiner resisted. "It is bad for AutoImmune to be on the West Coast," Weiner noted in his journal. "My input is too crucial." In the end, the board saw things his way. Weiner concluded that he had "dodged a bullet."

And so, after a headhunter found Bishop out in California, one of the very first questions Barry Weinberg asked him was whether he would be willing to move to Boston. "I told him," Bishop remembers, "if the opportunity's right I'll move to Nome."

Bishop is a big, robust man with a red face and a laugh that begins as a wheeze and spreads to his belly. In his late forties, he retains the full head of light-brown, wavy hair of his youth. Bishop's father was a salesman, with what he describes as a "marketing personality, very, very outgoing, opinionated" and given to quitting jobs when the boss didn't see things his way. The result was that the family fortunes fluctuated wildly.

Bishop resolved to provide his family with a greater sense of security. And, at least until he took the AutoImmune job, he had succeeded. Some years after completing his Ph.D. in biology, he supplemented it with a master's in business. Although his résumé reflects the merging and acquiring mode of the drug industry in the eighties and nineties, his trajectory, as he moved from one job to another, tilted upward. By the time he came to the attention of the AutoImmune headhunter, Bishop was running a $350 million pharmaceutical business for a drug company called Allergan. But he was aware that the president of Allergan favored the other "lead dog" over him for the job of chairman of the company, so he was willing to make a move.

One of Bishop's colleagues has described him as "a salesman disguised as a Ph.D." But in fact the salesman is the part of Bob Bishop that is obvious from the start. While he was completing his Ph.D., he worked for Baxter Pharmaceuticals one week out of every month, flying around to give sales talks about their products. He says even now that "talking is what I do best." He delivers speeches from the diaphragm, in a big round baritone with a hint of a cowboy accent. Only after a while does it become apparent that Bishop's "hail fellow well met" style masks a quick intelligence. Unexpected words like "truncated" pop up in his presentations. At the same time he was making sales pitches for Baxter, Bishop was helping to support his wife and young family by playing professional tournament bridge.

As a boy, Bob Bishop dreamed of becoming a doctor. But when he got to Berkeley, the lure of fraternity life and football wreaked havoc with his grades, and he was never able to recover enough to get into medical school. Working in the drug industry seemed like a natural alternative. Although he'd had opportunities before to follow in his dad's footsteps and "take a flyer," AutoImmune was the first start-up company that really tempted him. It was risky, of course, but he was impressed with the science. And, as everyone who has ever worked with him will attest, Bob Bishop is an optimist. This, as much as anything else, made him a sympathetic partner for Howard Weiner.

The night after he met and lunched with Bob Bishop, Howard Weiner wrote in his journal, "This could be it." Soon after that, Susie Bishop, Bob's wife, paid a visit to Boston to look for a house. In Wellesley, a western suburb, she found a Georgian brick house with a circular front drive and elegant grounds, including a garden pool with a fountain in the form of a fish. An experienced decorator of houses (the Bishops had bought and sold six already), Susie Bishop was excited by the house's possibilities. She came back to California and told her husband, "if you buy that house, I'll go." Bob Bishop agreed to buy it, sight unseen. Soon after, he took over as CEO of AutoImmune.

The Virtual Company Gets Real

One of the paradoxes of the drug business is that success in a big company can produce a yearning for something smaller. That was what happened in the case of Fred Bader, a Ph.D. in chemical engineering who spent ten years working for two pharmaceutical giants, first Upjohn and then Bristol-Meyers. The big companies moved slowly and he had a hunger for "cutting-edge" science. So he took a job at a fledgling biotech company, the Genetics Institute. It was "a wild and woolly company," Bader recalls, and it gave him "a real rocket ride" into executive responsibility. When he came, the company had 230 employes. By the time he left, the company had doubled in size and half of it "was reporting to me." To his dismay, he found himself "spending time with the other vice presidents talking about what we think the scientists are doing."

Once again, Fred Bader began to long for something smaller, where he could be in closer touch with the science. That was when he heard about "this virtual company where they were feeding chicken parts to people or something like that. It sounded a little strange." Bader decided to look into it. He met first with Bob Bishop, who was conducting interviews with prospective officers for AutoImmune, Inc. in the lounge of the Marriott Hotel in Cambridge.

Then he met with David Hafler and Howard Weiner. "You need to know if the scientists are good," he explains. When he met Weiner, the whole idea "began to gel in my mind. I was turned on to the technology and I was very turned on to Howard Weiner." Bader also liked the idea that "this is a platform technology. If this works there are so many other applications. Until now these diseases—diabetes, rheumatoid arthritis, MS—were studied separately, on the basis of their symptoms. Here was an opportunity to be in the forefront, and to study fundamental questions across diseases."

Fred Bader taught at the University of Michigan for three years before deciding he wanted to "get out into the real world." Tall and slight, he has a quiet thoughtfulness that suits a professor. He claims he almost lost out on a Ph.D. because he couldn't fulfill the foreign language requirement. But in English, he has a knack for the apt analogy. The drug discovery business, he says, is "sort of like prospecting. You dig a hole in the ground, you don't see anything. You keep digging, you might be two inches away from a vein of gold and walk away. On the other hand you could dig from here to China and never find a darn thing."

For Bader, signing on with AutoImmune meant coming to terms with who he wanted to be. "I figured out that I could be president of a biotech company, but it wouldn't be one I'd want to be president of." But was he right for AutoImmune? "Some of it was working up my nerve, having the confidence that I have something to offer at that early level." Bader decided to take the chance. He became AutoImmune's second employee after Bishop, with the title Vice President, Operations.

Not long after Fred Bader signed on, Malcolm Fletcher visited Boston at the Bishops' invitation. He was discontented with his job in Canada, where he was setting up a small research consulting company, and was looking for other options. Fletcher is an Englishman who defies the stereotype of British reserve. With his round glasses and a boyish white winning smile, he sets out to charm everyone around him and often succeeds. If he has a fault, it is that he talks too much, but often he is funny enough to be forgiven. He will tell you, for instance, that he is related to the great Alexander Fleming, who discovered penicillin, but will quickly add that the family didn't talk about the connection because Sir Alex was a lush. And he is likely to tell you, at your first or second meeting, that his contentious divorce, plus a new wife and family, have left him with big debts that a regular salary will never pay off. If AutoImmune succeeded, it could be a solution to his money woes.

The executive compensation at the new AutoImmune, Inc. would be comparatively low. CEO Bob Bishop was making $255,000, a 46 percent cut from his salary at Allergan, and Malcolm Fletcher, if he signed on as a vice president for Clinical and Regulatory Affairs, would make $125,000 to start. But in addition to their salaries, the top executives of the company would be given options to purchase stock in the company over time for next to nothing ($1.33 a share). Fletcher, for instance, would be able to purchase 75,000 of such options. If the company succeeded, the stock could go to $30, $40, or $50 a share. Then his troubles would be over.

Malcolm Fletcher comes from a British medical dynasty that, besides Alexander Fleming, includes a grandfather, father, mother, brother, and numerous cousins who were doctors. Fletcher, after training in Britain and practicing medicine for five years in rural western Canada, chose to follow in the footsteps of his most famous relative and go into drug development. He took a job at Burroughs Wellcome in Montreal, where he devoted his energies to designing and overseeing clinical trials of potential new drugs. It's a job that, according to Fletcher, is "a bit like doing a movie. You've got to get all these people together, you've got to build enthusiasm, you've got to deal with Murphy's Law, you've got to deal with disasters, you've got to establish a rhythm. You've then got to fuss people to death, because not everybody's good. And you've got to make it happen."

When he came to Boston, Fletcher met with Howard Weiner, and found him to have "the same single-mindedness, and the same ability to turn people on" that he'd observed in other successful innovators. Then Weiner sat him down in his office and left him alone to look at some raw, unanalyzed data from the first human trial of oral tolerance in rheumatoid arthritis, a trial in which patients were taking drops of collagen in their morning orange juice. Until that moment, Fletcher viewed oral tolerance as "an off-the-wall idea." But as he looked over the patient data, he suddenly got "this feeling of the hairs going up on the back of my neck." For the first time, he saw something that told him "there's biology there." He decided to take the job.

A Local Habitation and a Name

You could see the Ledgemont Center in Lexington, where AutoImmune Inc., as the company was now finally named, was setting up its offices, from Route 128, the beltway that encircles Boston. It was a pile of horizontal planes of red brick, interspersed with black glass. Its flat top was ornamented with an unruly array of skinny smokestacks venting the various chemicals cooking inside, in the labs of the small scientific companies that were its usual tenants. Closer in, it had a more elegant look. What had become the Ledgemont Center was once a gracious mansion, in the style of a Cotswold cottage, and the house had been ingeniously preserved as an anchor in the middle of the sprawl of newer brick. The old grounds, too, had been kept alive, and big pines and rhododendrons cloistered the center from the roads and highways around it.

In December 1992, Howard Weiner drove out to the Ledgemont Center and entered the new offices of AutoImmune, Inc. for the first time. Predictably, he took along his video camera, and shot pictures of Malcolm Fletcher, Fred Bader, and Bob Bishop behind their new desks.

The S Word

Howard Weiner has a favorite story he tells when he wants to explain the meaning of "artifact" in scientific research. A certain researcher who was trying to understand the causes of mental illness took samples of the urine of schizophrenic patients at a hospital in Salt Lake City, Utah. The researcher also took samples of urine from the hospital staff, to see if he could detect any difference between the two groups. Sure enough, there were certain chemicals in the urine of the patients that were not present in the staff's urine. A eureka moment! But it turned out to be a false eureka moment. Other laboratories in New York, Chicago, and California couldn't duplicate the Utah researcher's results. Although the patients came from diverse backgrounds, the staff of the hospital in Salt Lake were Mormons, and Mormons don't drink coffee. The finding was an "artifact" that had to do with coffee, not schizophrenia.

There were plenty of serious scientists who suspected that Howard Weiner's oral tolerance findings might be an "artifact" of some kind. Caroline Whitacre, the first person to demonstrate oral tolerance in EAE rats, was having a lot of trouble producing the active suppression that Weiner seemed to get with his animals. When Weiner presented his work in Denver, his old benchmate John Moorhead was incredulous about how feeding a fragment of the protein involved in EAE could possibly work to suppress it.

Anyone who claimed to be producing "suppression" was suspect in immunological circles, in any case, because of the word's ignominious history. Back in the 1970s, when cellular immunology was in its infancy, there had been a great deal of excitement about the existence of a "suppressor cell." Immunologist Jacques Chiller, who did pioneering studies on tolerance in the early seventies, remembers that the tools then available for studying T cells were inexact. Yet "all of a sudden people started describing an experiment that was very exact" and that posited a unique T cell with a suppressor function. "Everybody got caught up in this," Chiller remembers. One of the leading believers, Richard Gershon at Yale, had a cartoon in his office of an "immunological orchestra." There were a cluster of immune cells, each playing its own instrument, and the whole ensemble was conducted by a "suppressor cell." So great was the enthusiasm for suppressor cells that Chiller, who had been unable to find them in his lab, was booed at an immunological conference in 1974 when he tossed off a joke that cast doubt on the idea.

"Immunologists are very imaginative," says Chiller, "you have to be—particularly in the old days because it was so goddamn confusing. I blame some of these things on a combination of inexact science and overzealous creativity, interpreting a dime's worth of data into a million dollars' worth of theory." When molecular science got more exact, the suppressor cell idea died a sudden death. Its executioners were molecular biologists at Cal Tech, who cloned the immune genes of the mouse and found that there were no "suppressor" cell genes in the place where they were supposed to be located. "Then," Chiller remembers, "it was like that *Saturday Night Live* skit: 'Never mind.'" The suppressor cell idea "just disappeared overnight. Nobody said, 'Oh, it must be somewhere else.' Gone." Vijay Kuchroo, an expert in autoimmunity at the Weiner lab who earlier participated in the search for a suppressor cell, remembers that "suppressor" became the "S word" among immunologists. Now Weiner's lab was daring to use the S word once again, with Kuchroo as one of the senior advisors in the undertaking.

Adding to the skepticism of other scientists was what seemed to some like a simple-minded approach to complicated autoimmune processes that were not completely understood. No one knew, for instance, exactly which proteins were attacked in the joints of rheumatoid arthritis sufferers. Yet, according to the logic of oral tolerance, you would need to feed the patient the protein that was involved in their arthritis to activate the T cells in the gut, which would recognize and attach to the corresponding tissue in the joints. Why choose to feed collagen, for instance, when it wasn't necessarily the protein attacked in the synovial tissue in the joint, causing rheumatoid arthritis (RA)?

Similarly, more than one myelin protein was involved in the attack on the myelin sheath that caused multiple sclerosis. Around 1990, researchers in California described a process they called "spreading autoimmunity" or "epitope spreading," which occurs in MS over time. Early inflammation may be caused by myelin basic protein (MBP), one of the myelin proteins that is most abundant in the myelin sheath. But later on, other proteins become involved in the autoimmune inflammation, including PLP (proteolipid protein) and MOG (myelin oligodendocyte glycoprotein).

The question was, how could feeding one protein—collagen in the case of rheumatoid arthritis, and MBP in the case of MS—shut down autoimmune activity that involved multiple proteins? Until this question was answered, there were going to be many more doubters than believers in immunological circles.

In 1990 and 1991, when the first human trials in oral tolerance and MS were going on, Ariel Miller, the Israeli successor to Ofer Lider in Howard Weiner's lab, came up with an answer to this question, at least as it applied to laboratory animals. He knew that Lewis rats, when fed MBP, were resistant to EAE, the animal model of MS. He knew too that he could extract T cells from the spleens of the resistant rats and inject them into other animals, who would then be resistant as well. How did those T cells manage to transfer suppression? In order to understand the mechanism, he needed to take a closer look at them.

So Miller extracted those particular T cells from the spleens of his resistant rats and put them into a transwell, a container divided horizontally by a thin membrane that would permit the passage of soluble substances but that cells were too large to penetrate. He wanted to find out whether or not the T cells on one side of the membrane had to be in contact with cells on the other side to suppress their proliferation. It turned out that they didn't. The T cells were releasing a soluble factor that could cross the thin membrane and suppress.

This discovery led to a hypothesis about the *S* word that was both more complicated and more plausible. The T cell involved wasn't a "suppressor cell," but rather an agent of suppression. Under the right circumstances, when stimulated by oral antigen, the T cell released a "suppressive factor" that crossed the transwell membrane and did the job. Not only that, the suppressive factor wasn't choosy: It suppressed not just cells reactive with MBP but bystanders in the vicinity, using, as Miller's paper on the subject concluded, an "as yet unidentified antigen-nonspecific mechanism." Weiner and his group gave this phenomenon the name "bystander suppression."

Jacques Chiller, who had rightly criticized the "suppressor cell" when it came into vogue in the seventies, was skeptical about the work of the Weiner lab as well. But when he heard Weiner describe "bystander suppression," he changed his view. The earlier idea was that there was a "a cell of a different lineage" that suppressed. That turned out to be wrong. But the idea that regular T cells, responding to fed antigen in the right environment, switch on certain sets of genes that release suppressive factors, or cytokines, seemed "viable." "Is it wholly real, is it proved?" asks Chiller. "Of course not. But is it practical, can you take advantage of it? I'm one of those who says, 'let the experiment be done.'"

CHAPTER 8

The Road Show

BOB BISHOP AND HOWARD WEINER were well aware that scientific founders of biotech companies often clash with the "suits" in management, sometimes with disastrous results. So right from the start, the two made a public display of liking each other. They called each other "pardner" and vowed to answer each other's phone calls, even if it meant interrupting a meeting. Bob Bishop got copies of the sayings Howard Weiner had on the wall in his office and put them up on a wall at AutoImmune, Inc. The two hung posters in their offices by Mordillo, showing two cartoon golfers in a tight spot. One golfer is at the edge of a cliff, holding onto a rope that is tied around the waist of the other, who is dangling over the chasm. There is a golf ball caught in the crook of a scraggly tree sticking out from the side of the cliff. The suspended golfer is taking a swing at the ball, while his partner hangs onto him for dear life. In Bob Bishop's picture, the golfer on the cliff holding the rope is labeled "Howard" and the one in midair, taking the nearly impossible shot, is labeled "Bob." In Howard Weiner's poster, it's Bob who is holding the rope while Howard takes the shot.

It's possible that the requirement that they act like friends gave Weiner and Bishop, two hard workers with little time to cultivate male companionship, permission actually to become friends. Certainly, as the months went by, they came to thoroughly enjoy their time together. Whenever they had a chance, they played golf; they even went to Bermuda together for three days of golf and brainstorming. They ate out and went to plays with their wives, so the couples could get to know each other. But for both of them, the most exciting part of their relationship had to do with building the company. The most intense period in that process had to be the three whirlwind weeks of fund-raising called the road show.

Around the time that Bob Bishop signed on, the board of AutoImmune decided that the next step for the company would be a large, pivotal phase 3 trial of oral tolerance as a treatment for multiple sclerosis. This was an unorthodox decision. In general, human trials proceed through three stages. Weiner's group had already conducted a phase 1 trial at the Brigham, meant to establish safety. But the intermediate step, a phase 2 trial of perhaps 100 to 300 patients to evaluate dosage and effectiveness, was going to be passed over. Phase 3 trials are the make-or-break trials, designed to determine whether the drug has an effect that is statistically significant. Even though the initial trial of oral myelin in thirty MS patients had produced promising results, they were not statistically significant. It was a very small trial on which to base a large, phase 3 trial of the drug.

The reasoning, in the words of Barry Weinberg, was "we saw the ability to actually get this technology into humans very early at a very low cost. In most cases in the pharmaceutical industry it's a very very long path to get drugs into humans because of safety issues. Here there was absolutely no indication there would ever be safety problems. Furthermore, we had a disease, MS, for which there was no treatment, and the ability to get compassionate approval of early stage trials was a lot easier."

But the trial, which would last two years and involve hundreds of multiple sclerosis patients at many sites in the United States and Canada, was going to be extremely costly. AutoImmune, which had eight employees at the end of 1992, was going to have to expand exponentially. The company would need to hire in-house scientists and outside consultants to help administer the trial. There would be expensive meetings of the doctors and administrators from all the participating sites, and all the investigators would have to be compensated for their participation over the two-year period of the trial. The trial drug would have to be prepared and tested, according to the stringent standards of the FDA. Every one of the 500 patients in the trial would undergo extensive testing. Further, every single step in the process would have to be documented in great detail. Typically, when a company applies to the FDA for approval of a drug, the documents describing the clinical trials run to at least 100,000 pages.

What's more, AutoImmune planned to conduct several phase 2 trials simultaneously in rheumatoid arthritis and to underwrite research in three other areas—diabetes, uveitis, and organ transplant. All told, Bishop told the AutoImmune board, he was going to need about $50 million to carry out the company's ambitious plans for the next three or four years. At that point, in late 1992, there was $3.8 million on hand. It was time to raise money by taking the company public.

Bishop had begun to talk about an initial public offering (IPO) with investment bankers as soon as he took the job. They told him he needed to hire officers. "They said we can't take it public with just you, as much as we like you. You gotta have some company here!" Now that Bishop had hired Bader and Fletcher, as well as a chief financial officer named Tom Hennessey, the bankers were more receptive. In exchange for a percentage of the money raised, two San Francisco investment firms, Hambrecht and Quist and Montgomery Securities, agreed to organize a road show for AutoImmune. The road show, a grueling multicity tour in which the company's principals make their case to a group of invited institutional investors, analysts, and money managers, is essential to a successful IPO launch.

Before the road show, Bishop and his team put together a prospectus, a documentation of the company's history, financial situation, and goals. The prospectus, which has to be approved by the Securities and Exchange Commission, is the document that is handed out to potential investors during the road show.

Montgomery and Hambrecht and Quist, veteran planners of such launches, then laid out an itinerary for Bishop and Weiner that took them zigging and zagging across the United States and Europe, meeting with single investors and with groups, large and small. During eighteen days of travel, Weiner and Bishop would work as a team, giving their presentation and answering questions in thirty-seven separate meetings.

One secret of success, on a road show, is to make investors feel that they have exclusive access to new information. Thus, the meetings are off limits to all but the chosen few. Nonetheless, somewhat to the discomfort of Barry Weinberg, Howard Weiner brought his video camera along. Just before he and Bishop left, on Sunday, January 3, 1993, he filmed a run-through of the twenty-five-minute presentation at his house in Brookline.

A newcomer to the biotech sector, listening to the Bishop and Weiner presentations, might well come away asking, "where do I sign up?" The slides were simple and clear, the science was presented in a way that made the layman feel intelligent, and the concept looked, as a number of audience members would remark during the tour, almost "too good to be true." The presenters were impressive too. Weiner, the Harvard researcher, had already explained oral tolerance many times and did it with ease and conviction. "You could eat dirt and not get sick," he would say provocatively, as he explained the special properties of the gut immune system. Weiner was also the good doctor—a reference to "sitting with patients as a clinician" was artfully inserted into his discussion. Investors, like everyone else, tend to admire a top doctor in the field. In fact, several times during the road show, individuals came up to Weiner after the presentation and asked if he would see someone in their family with MS, or at least render an opinion.

Bishop, a little less practiced at this early stage, had a voice that exuded confidence. Even when he mentioned that something might not work out, as for instance in the case of a partnership between Schering-Plough and AutoImmune, the listener got the impression that he was only bringing it up out of an excess of caution.

Bishop started the presentation rolling with "a brief review of the immune system," less intimidating to businessmen coming from one of their own. But very rapidly he moved on to what mattered most to investors. The "technology" in question, he pointed out, "addresses some very big markets." Bishop used the phrase "big markets" three times in his presentation—first briefly, tantalizingly, before he yielded to Howard, then in more detail when he returned. There are about 500,000 MS patients worldwide, 7 million rheumatoid arthritis patients, 100,000 patients with the eye disease uveitis. The current market for autoimmune drugs in the United States was $4 billion and was expected to double in the next few years. Oral tolerance had the potential to be applied to a host of autoimmune diseases—here Bishop flipped to the slide that listed many illnesses, including others that might be familiar to the lay investor, such as psoriasis and myasthenia gravis.

All this was impressive enough. But there was more. Everyone knows that side effects are a major problem in developing and marketing drugs. And oral tolerance had no side effects! What's more, this drug, unlike others in development for MS, was taken by mouth. You don't have to be a scientist to know that people who will refuse an injection are willing to take a pill. There seemed to be the promise of a relatively fast return on investment here too. Bishop was talking about taking the first of the oral tolerance drugs to market in just three years. Who wouldn't want to bet the farm?

A more experienced investor might notice some things that would give him pause. First and foremost would be the small number of humans in the trials that Bishop and Weiner were touting—thirty patients in the only human MS trial, ten in the only completed arthritis trial, and only two so far in a very preliminary uveitis trial. It was also worth noting that Bishop and Weiner were taking a leap that is not uncommon in small biotech companies: They had blended the usual phase 1 safety trial and phase 2 efficacy trial into what they called a phase 1/2 trial, and a small one at that.

A sophisticated investor would also know how hard it is to succeed with an experimental drug. Only one in five that make it to human trials are approved by the FDA. And it takes a long time: On average, it takes fifteen years for an experimental drug to travel from the lab to the medicine chest, according to the Tufts Center for the Study of Drug Development. In that context, Bishop's statement that "we hope to get our first products to market in 1996" might raise a few eyebrows. The idea that you could begin a two-year trial with rolling admissions in 1993, then finish in time to get approval from the FDA, a process that takes at least six months and averages nineteen months, and get a drug on the market in 1996 would look wildly optimistic to an investor who knew the drug industry. An old hand would suspect that Bishop's projected "burn rate" for the company of $550,000 a month would be likely to continue considerably beyond 1996.

On the opposite side of the ledger would be the fact that the animal trials of oral tolerance had been successful, as Bishop pointed out, "in three diseases." "Each on its own is impressive," he noted. "Combined, they make a convincing case for our therapeutic approach." Furthermore, Weiner dropped in a mention of the fact that the results of the first human MS trial were about to be published in one of the top scientific journals in the world, *Science*. This was a strong indication that his peers respected the science. Also, the company was using what Bishop referred to as a "belt and suspenders" approach to developing the drugs: a fast track for the MS drug, and a more deliberate approach with multiple phase 2 trials for the rheumatoid arthritis drug. This would be somewhat reassuring. The safety profile of oral tolerance would impress knowledgeable investors even more than naive ones: The toxicity of drugs is the most important single barrier to success in human trials.

Whether experienced or not, all the investors listening to the Weiner and Bishop presentation knew that an investment in AutoImmune was highly risky, but offered the possibility of high gain. The conversations that then took place in the privacy of various boardrooms and hotel rooms around the world convinced many that AutoImmune's potential made it worth betting on. When it was all over, Weiner described the trip as "exhilarating," adding "it was wonderful for Bob and I to be together."

The Weiners' house is strikingly modest: a two-story brick and clapboard house with four rooms on the ground floor, including Howard's cramped study. The kitchen cupboards have a pseudo bleached barnboard finish. Most of the backyard is taken up by an asphalt driveway and an oak tree. It was a far cry from the lavish settings Bishop and Weiner encountered on the road trip.

Because the actual meetings with investors were off limits, Howard Weiner's video of the road trip is a fragmented record of the beautiful places where the rich, and their representatives, do business. In between numerous shots of Bishop, Weiner, and others asleep on planes or talking in the backs of limos, there are shots of empty meeting rooms on very high floors of fancy office buildings. Sometimes, the tables are covered in linen and laid for breakfast or lunch, with an AutoImmune prospectus on top of each plate. Other times, the tables are littered with dirty dishes, postmeeting. There are also shots of Bishop and Weiner's various opulent hotel rooms.

For the most part, Bishop and Weiner seemed to revel in the luxury. But from time to time, there were hints that Weiner was a little uneasy with pitching his ideas for big bucks. When Bishop asked him, on the first day, to predict their future, Weiner said, "I think we're going to raise so much money that I'm just gonna vomit!"

Bishop, laughing, responded, "Is that what you do when you raise a lot of money, Howard?"

Weiner shifted to a sports metaphor. "No, we're just going to get men on base and bring 'em on in."

In New York, many meetings later, Bishop and Weiner were going up to a high floor in the Helmsley building for yet another presentation. "Leona Helmsley," Weiner narrated as the camera panned the rococo interior of the elevator, "now *she* made a lot of money. But then she went to prison!"

Most of the time, though, Weiner was as upbeat and boosterish about the mission as Bishop. On the first day, they filmed each other walking toward the TransAmerica building "because that's where they keep the money." The rather lame joke about going to the various places "because that's where they keep the money" was carried on through the trip—through visits to Stockholm and Copenhagen, through the gourmet lunch in Paris, where they were pleased that the French actually stayed awake through their presentation, and through numerous pitches to small and large groups in London and the beautiful surrounding countryside.

On the plane home, after the European tour, Weiner and Bishop felt triumphant. As they filled out the customs forms, asking how much money they were bringing into the United States, Weiner said to Bishop, "we're bringing back millions." Apparently someone on the plane overheard this boast and tipped off the customs inspectors. Bishop was thoroughly searched from head to toe at Logan Airport before they were allowed to go home to their wives.

Back in the United States, the road show returned to San Francisco, then moved on to Dallas, Minneapolis, Boston, and finally New York because, as Weiner and Bishop noted on the video, "New York is where they *really* keep the money." By that time, the "order book," in which investors commit to buying a certain number of shares, was building so well that Hambrecht and Quist suggested increasing the number of shares to be offered to the public.

The bankers running the road show make the rather arbitrary decision about how many shares to offer, based on past experience and their opinion of the company's potential. If things are looking good, it is in the bank's interest to increase the shares offered, since their profit is a percentage of the money taken in at the initial public offering. (In the case of the AutoImmune IPO, Montgomery and Hambrecht and Quist's combined percentage was about 7 percent.) The company, however, doesn't want to offer so many shares that they limit their ability to come back to investors a second time later on.

Because the road show was going well, David McCallum, the head of health care investment banking for Hambrecht and Quist, suggested, at a meeting with Weiner, Bishop, and Weinberg in H&Q's New York offices, that they increase the initial public offering from 2.5 million to 3.5 million shares. But the AutoImmune team was wary.

"We were debating how big we wanted to make the deal," Bishop explains. "And we actually asked them [the Hambrecht and Quist advisers] to leave the room so we could have this discussion and arrive at where we wanted to be. Howard went to the board and began to sketch out what the options were and what the levels of cash were that we'd have back in the company as a result."

"These discussions take time," Weiner notes. "They wanted us to issue a million more shares—obviously they're making more money if they issue those shares. The order book was building up, the order book was ten times oversold, we don't know what's going to happen to the market. So their recommendation was to increase it a million." In the end, the AutoImmune team decided to increase the offering by 20 percent, to 3 million shares. The price per share was yet to be finalized, but all agreed it would probably fall between $11 and $13.

The eighteenth and final day of the road show, January 20, 1993, was also the day of Bill Clinton's inauguration to his first presidential term. The day before, Centocor, another biotech company, had crashed, and the uncertainty introduced by Hillary Clinton's plans to initiate universal health care coverage was already dulling enthusiasm for investment in the biotech sector. AutoImmune got its offering out just in time.

Bishop and Weiner had devised a scheme to measure their success. If they raised only $25 million, they would give themselves a barely passing grade, a D. Raising $29 million would be a C, $32 million a B, and $35 million an A. They raised $39 million. "So we got an A+," Bishop says. Because the road show had gone so well, the AutoImmune team pushed to raise the share price to $14. But the bankers, wanting a good deal for their regular investors, resisted. In the end, Bishop agreed to put the stock on the market at $13 per share.

On the night of January 20, Bob Bishop delivered Howard Weiner back to Mira and his small brick house in Brookline. With Mira filming, the two performed a final ritual: handshake, followed by handclasp, followed by a bear hug. Then they faced the camera, with arms around each other's shoulders, and gave a simultaneous thumbs-up.

A month later, the report on the first human trial of oral myelin in MS came out in *Science*. Even though the authors pointed out that "conclusions about efficacy cannot be drawn," the paper got an immediate press reaction. There were reports on CNN, NPR, and the BBC. The *New York Times* Science section ran an article headlined "Multiple Sclerosis Vaccine Shows Early Promise," complete with a diagram tracing the myelin capsule's journey into the gut, and the regulatory T cells' journey out to the myelin sheath. A telephone receptionist at the Brigham MS Center, hired in anticipation of the publication in *Science,* was swamped with calls from hopeful patients. "It is the coming of age of Oral Tolerance," Weiner wrote in his journal, but added that "much will depend on the arthritis and uveitis work."

A month after the *Science* article appeared, the sixty-patient phase 2 trial of oral collagen as a treatment for rheumatoid arthritis was unblinded. Once again, there was good news. Of the twenty-eight patients who were taking the collagen drops each morning, four had a complete remission and most of the others had significantly less joint pain, swelling, and tenderness. None of the patients on placebo had full remission, but four experienced a reduction in symptoms.

In September 1993, the results of the second oral tolerance trial also appeared in *Science,* with David Trentham as first author, his wife Roselynn second, and Howard Weiner last.

Once again, the press picked up the story. The *Boston Globe,* in an article in the Metro section, chose to focus on a fifty-seven-year-old rehabilitation nurse named Jean Arns, who was diagnosed with rheumatoid arthritis at age forty-four. Her pain had become so severe that she had trouble opening bottles of medication for patients, or even moving from room to room to do her job. Before the collagen trial, she had tried one medication after another and found little relief. After she began to take collagen her symptoms diminished dramatically. She could "walk up the stairs without holding onto the handrail." She could even go bicycle riding.

Patients responded in overwhelming numbers to the news. There were so many phone calls to the Beth Israel Hospital, where Trentham and his group were based, that it tied up lines and caused resentment. Some in the rheumatology community were angry as well. They criticized the publication of the paper in *Science,* a theoretical journal that isn't read by most clinicians. If it had appeared in *Arthritis and Rheumatism,* rheumatologists wouldn't have been blindsided by patients demanding treatment. There was also a feeling that it was pretentious to publish the results in a premier scientific journal, rather than a clinical journal. But Weiner and other authors of the paper defended their choice of *Science.* In the words of Lea Sewell, another rheumatologist whose name was on the paper, the trial was a test of "an immunological principle, which is that you can actually build up tolerance by having this low dose of protein arrive in your gut and having the immune system in the gut look at it every day." That principle was more central to the paper than the promise of a treatment.

The oral tolerance principle laid out in the two *Science* papers has indeed proved to be important in the field of immunology. When the papers were published in 1993, there were only a handful of people working on oral tolerance. But at the 1999 conference of FASEB (Federation of American Societies for Experimental Biology), a meeting of 10,000 biologists from all over the world, there were at least three dozen papers on oral tolerance. Over 100 papers a year are now published on the subject.

Despite public overreaction to the *Science* papers and fallout among colleagues, the year 1993 was one of scientific triumph for Howard Weiner. It was a banner year for AutoImmune, Inc., as well. But it was not without setbacks. Novo Nordisk, the Danish company Bishop and Weiner had been courting as a potential partner, pulled back from making a commitment. And, despite the promising results of the arthritis study, Schering-Plough, which had already paid a $500,000 option fee to AutoImmune and was scheduled to spend millions more, decided not to renew the agreement it had entered into in 1992.

Bob Bishop told Weiner the disappointing news about Schering-Plough when they met to play golf one Saturday in May. As they walked the course, the two of them tried to figure out what went wrong. The agreement with Schering-Plough was "a very rich deal," Bishop knew. In the year after the deal was signed, pressures had increased on pharmaceutical companies to lower prices, partly owing to what Bishop called "the Hillary factor"—Hillary Clinton's attempt to pass national health care legislation. So the oral tolerance project, when compared to internal projects, may have lost out. Bishop guessed they may have also been up against a "not invented here" syndrome. Internal projects had an edge. Or perhaps, a thought both Weiner and Bishop resisted, Schering-Plough didn't think AutoImmune had enough evidence of efficacy in humans.

As a result of the announcement that Schering-Plough had pulled out of the deal, AutoImmune's stock price, which had risen at one point to $15 a share, dropped to $5.50 and stayed low for months. A consultant from a public relations firm told an AutoImmune board meeting that people on Wall Street didn't believe in the oral tolerance technology. They saw it as "one man's pet theory." Around the same time, Betaseron, the first of three injectible drugs that were in the pipeline, came on the market and became the first drug available explicitly for the treatment of MS. Even though there was hope of an agreement between AutoImmune and Eli Lilly, which had expressed an interest in collaborating on diabetes trials, nothing had come through yet.

None of this, however, got in the way of plans to launch a phase 3 trial in MS of oral myelin. In October, Howard Weiner, Bob Bishop, Fred Bader, and Malcolm Fletcher, along with several new AutoImmune employees, convened at a Holiday Inn in Washington, D.C., to prepare for a meeting with the FDA, in which they would present their plans for clinical trials over the next several years.

At this point, the oral myelin had been given the trade name Myloral (a combination of myelin and oral). Myloral would be tested on about 500 patients with relapsing and remitting multiple sclerosis at fourteen sites around the United States and Canada in a double-blind study over a two-year period. Myloral was to be AutoImmune's "lead product," the one they would try to get to market as quickly as possible.

AutoImmune's rheumatoid arthritis drug was assigned the name Colloral (collagen plus oral) and placed on a slower track. Colloral would be tested in not one but a series of phase 2 trials. If the phase 3 Myloral trial was disappointing, there would still be resources to go into phase 3 trials in Colloral. The Colloral phase 3 trials would be built on the more solid ground of numerous preliminary studies. The FDA allowed the company plan to go forward.

Six months later, the neurologists and study coordinators from around the United States and Canada who had been recruited by AutoImmune to conduct the phase 3 Myloral trial gathered for a two-day orientation at the Harbor Hotel, near the airport in Boston. And on March 30, 1994, Alex Rae-Grant, a neurologist in Allentown, Pennsylvania, enrolled the first MS patient in the study. A decade after Howard Weiner got the idea of feeding myelin to mice, a phase 3 trial in humans was under way.

CHAPTER 9

The Investigators

Lehigh Valley, Pennsylvania

For the most part, Malcolm Fletcher and Jim Parmentier, AutoImmune's newly appointed clinical trials manager, chose academic centers in metropolitan areas as sites for the Myloral trials. But there were a few choices that resulted from special pleadings of one kind or another. Barbara Hastings, an elegant blonde neurologist who is in private practice with her husband in Tulsa, Oklahoma, was excited about the "brilliant but simple" idea of oral tolerance when she heard Howard Weiner speak about it at an academy meeting. When she got home to Tulsa she faxed Weiner and asked if she could participate. Fletcher and Parmenter had their doubts; Hastings had only a part-time connection to the university and limited facilities for testing. "It took me six months to convince them," Hastings says. "I told them, 'You need me epidemiologically. You need me because of where I am in the latitude, just a little below the meridian, and because of the rural drawing power.'" Finally, after Fletcher paid a visit, AutoImmune agreed to include the Tulsa site in the trial.

Another atypical site was the Lehigh Valley Hospital, which draws its patients from a 100-mile radius in the hilly region of eastern Pennsylvania at the edge of Appalachia. Allentown and Bethlehem, with its Moravian church and beautiful cobbled streets, are in the Lehigh Valley area. But so are towns like Red Hill, where everybody drives a pickup truck. Compared with Myloral trial sites in St. Louis, New York, Los Angeles, Montreal, Toronto, and Baltimore, Lehigh Valley didn't have a very large population to draw on.

But Lehigh Valley had Alex Rae-Grant, a vigorous young neurologist who was enthusiastic about the Myloral study. His research group at Lehigh had participated in one of Weiner's earlier cyclophosphamide studies. "We didn't do a lot on that," Rae-Grant concedes, "but Howard came back and forth a few times and I went up there a couple times. So when the Myloral came up they were willing to look at us as a center, although we'd never had big MS trials before."

Alex Rae-Grant is something of an anomaly himself: Born in England to a family of Scots lineage and raised in Canada, he was a champion breaststroker at Yale before he enrolled in McMasters, a medical school in Hamilton, Ontario. Strappingly handsome, with receding blond, slightly curly hair, Rae-Grant recently appeared in the hospital's annual report wearing a bike helmet and shorts, on a tandem bicycle with his son. Father and son were participating in the Delaware Valley bike tour to raise money for MS.

When Fletcher and Parmenter came to visit Rae-Grant, he was ready for them. The Lehigh Center had already done a number of studies on stroke, and had research coordinators in place for future projects. But even more impressive was the pool of potential patients. "Well before we knew the study was going on, we had started developing a list. So we had about 150 people by the time they came down." Other centers might have as many patients, but quite a few were disqualified because they had participated in previous MS studies. In this way Lehigh Valley's inexperience with clinical trials was an advantage. As Hillel Panitch, who headed the Myloral study at the University of Maryland in Baltimore, noted with a hint of envy, "They had this whole stable of virgin untreated MS patients. They enrolled sixty-five patients in the time it took us to enroll thirty-five!" Lehigh Valley wound up enrolling more patients than any of the other fourteen sites in the study.

Rae-Grant sees trial patients in an anonymous brick building with spartan interiors, across the road from his office in the hospital. The atmosphere is less pressured, he says, than in his regular office.

"Since there's good evidence that people have fewer attacks and do better on placebo," Rae-Grant says, "I tell people at least it's better to be in a study than not in a study. I'm a firm believer that people do better if they think they're going to do better. There's good evidence that that affects immunology."

The placebo effect, however, although it may be beneficial to the patient, can present problems to researchers. If the study drug has an uneven effect, helping some and not others, its benefits can easily be masked by the placebo effect, particularly in an optimistic atmosphere like the one in Allentown.

Though he is hopeful that Myloral will work, Rae-Grant worries that the "primary endpoint" of the study, which is whether MS patients have fewer attacks over a two-year period, may not really get at the most damaging aspect of the disease. When you study MS patients in early stages, you can't look at the one thing that matters most—their degree of disability—because it hasn't really showed up in any dramatic way. And so you measure attack rate, even though that's not what is most important to the patients.

"If I could tell somebody they were going to have an attack of MS two or three times a year, which would make them sick for a week or two, and yet at the end of four years they would be no more disabled than they are now, they would say 'I'll take it.' If I told somebody, 'you're never going to have an attack but gradually over the next ten years you're going to become disabled and go into a wheelchair,' they'd say I'd rather have the attacks. What you really want to do is a study of the drug for twenty years. But nobody is going to fund that."

Rae-Grant has no way of knowing which of his patients are on the real thing and which are on placebo, and that makes it impossible to guess whether the Myloral is working. "We have some people who came in with definite MS, with two attacks within the two years before the study, who've just stopped having attacks and are functioning well. We have other people who have progressed to the point of significant disability, two or three have gone into wheelchairs during the study. So there is a spectrum."

If Myloral does work, he predicts that its effect will be modest. "Are we looking for it to stop disease? That's probably asking too much. Because one of the things about the study is we don't really know what the right dose is. We don't know if this should be used in combination with other medicines. But I would say we're looking for a 30 percent or more decrease in the number of attacks per year. Why that number? Well, that's the number that all the other three major agents (the ABC drugs) have shown. So if you can do as well or better than them with an oral agent, then you're off to the races."

Back at AutoImmune, Inc. in Lexington, Vice President Fred Bader and others have spent a lot of time, since the trial began, worrying about whether the "mad cow disease" (BSE) scare in Great Britain will make people reluctant to use Myloral made from cow brain. But in Lehigh Valley nobody seems concerned about the association. There's a Gary Larson cartoon in Rae-Grant's office, brought in by a patient, about a cow with an "oversized brain." And there are cows in various incarnations—stuffed, poured, and carved—in the offices of Nancy Eckert, the nurse coordinator of the trial. Nancy, a small but sturdy grandmother, is as responsible as anyone for the upbeat atmosphere around Lehigh Valley. "I'm mother Myloral," says Nancy, with her outsized throaty laugh.

"When I first started in the study," Nancy says, "issues mostly revolved around 'do I take my pills one time a day or four times a day?' . . . Then I got a little more into 'I have a cold . . . should I take a cough syrup or not?' That kind of thing." Sometimes, patients call now because they just need someone to talk to. And then, says Nancy, rolling her eyes, there is the problem of the "new voodoo medicine."

One patient in the study was going to see an "iritologist." "An iritologist is someone who . . . by looking in your eyes can tell you what's wrong with you and what kind of treatment you need." Another patient in the study was taking thirty-eight different types of vitamins and oral preparations a day, some of them four to six times a day. "Finally I said, 'Now look, your body will only accept so much. And if you feel you want to give this money away, here's a little envelope you can just stick it in. Give it to the study, we'll have a party!'"

Nancy Eckert, whose face is just beginning to take on the wrinkles and softness of an apple doll, has none of the cautious deference some nurses might feel toward a superior, the study doctor. She's blunt, to the point, and full of stories.

"I had three truck drivers in the study," Nancy says. "Three big cross-country eighteen-wheeler heavy haulers. One guy just decided that he couldn't handle it *at all* and he dropped out. And another fella is just as compliant as can be. I call him and tell him you gotta come in here and stand on your head for an hour, he will come in here and stand on his head for an hour. Then I have the other fella—oh my word, talk about Peck's bad boy! He crosses the country all the time. He thinks that he can stop at any neurologist and they will examine him for this study. And I could not make him understand that you take all the pills in one bottle and *then* go on to the other bottle."

Nancy's getting going now on noncompliance. "Getting some people to keep their diary! I have one guy, I call him the pristine book. His book is just as pristine as the minute I handed it to him—there's not a single solitary thing in it, nothing at all! And I say, 'Did you take any pills?' 'Oh yeah.' 'Well, did you take anything other than what I gave you?' 'Oh, I had a bad cold . . . I took

Nyquil.' He's had a cold, he cut his foot, he had a rash, he had a headache, he missed two days because he was sick to his stomach on his pills. It's tough."

The operative word here, as in every study, is control. If you can control the variables, you can produce credible results. And so there are identical notebooks, and patient diaries, and lists of questions to ask everyone and lists of procedures for every patient to follow. For every one of the 515 patients enrolled in the Myloral trial, there are two three-ring binders, covered in pearly plastic and labeled "A Double-blind Placebo-controlled, MultiCenter Study of Oral Myelin (Myloral) in the Treatment of Relapsing-remitting Multiple Sclerosis." Only the colors on the spines differ: The women patients get pink labels, the men get blue. "The best person for this job," as one study coordinator pointed out, "is an obsessive-compulsive."

Like Rae-Grant, Nancy Eckert is optimistic about Myloral. One patient told her, "If I am on the real thing, I don't have too much respect for it!" But a lot of patients have done well. Nancy even thinks Howard Weiner's oral tolerance breakthrough may win him a Nobel Prize. She told an MRI technician she knows to "hang on to that little black dress and those pearls, because we may be going to Stockholm!" Dr. Rae-Grant tells her, though, that it won't be for another twenty-five years. To which Nancy replies, "That's all right, you'll still be young enough to push me!"

Allison Hay was diagnosed with MS by Alex Rae-Grant when she was twenty-seven. She was recently married, and she and her husband Bruce had just bought a house and were about to start a family. When she got the diagnosis, "I just felt like the world came crashing down." On the one hand, she was relieved that the numbness that had taken over her body was not caused by a brain tumor. On the other hand she imagined that she would be in a wheelchair "very soon."

Since the diagnosis, Allison has been hospitalized with an episode of optic neuritis that caused her vision to blur. And she still gets so tired at times she can't function, especially in heat. Now her main complaint is spasticity in her legs that makes them feel "runny" at night. "Sometimes, I want to take 'em and hang 'em up on the back of the door," Allison says. "I really do." Allison Hay lives near Allentown, Pennsylvania now, but she speaks with the soft diphthongs of North Carolina, where she grew up.

A petite thirty-one-year-old with a perfect pageboy, Allison loves the phone, which makes her a natural for her job, processing claims at the Guardian Insurance Company. But she insists she'd rather stay home and be an old-fashioned wife. She and her husband Bruce have a conventional division of duties—she does the shopping and cooking, he pays the bills and takes care of finances. "I just want to create the bills," Allison says. Most of all, Allison wants to "be a mom." The hazards of childbearing with MS are what preoccupy her the most.

Ever since they got the diagnosis, Allison and her husband have been gluttons for information. "We checked out every book in the library, we bought books, and I called the MS Society right away." What they learned about pregnancy gave them pause. "Pregnancy is nine months when your body is actually very happy," Allison explains, "because of those hormones goin' through it." The problem comes after the baby's born. "That's when you're going to have an exacerbation, and it may be the worst one you've ever had."

At Allison and Bruce Hay's modest split-level in North Catasauqua, there is a sense of arrested motion. A huge formal photograph of Allison in her wedding gown hangs over the sofa, where the baby pictures might be. "That's my Princess Di dress," Allison says, as she shows me around. Except for Bruce's unadorned study, where computers and computer magazines cover all the surfaces, Allison's house is frilly—lace curtains, beribboned straw hats, lots of pillows, and a Danbury Mint doll collection in the spare room. Allison views the place as a "newlywed house," to be replaced later by something nicer. The current furniture is hand-me-down, and she's waiting until Bruce becomes a manager and they entertain before she unpacks the good china. Bruce, she says, is "quite a go-getter." He just got his MBA while holding down a fulltime job, and "he aspires to go out on his own one day as a consultant."

Allison has adorable good looks: a small face, but with a rather large mouth and a big, bright smile. She wears tiny diamond earrings, a diamond bracelet on her slender right wrist, and glasses with delicate filigree stems. Her clothes are classic—gray slacks, white turtleneck, red blazer—and her figure's perfect. It's not surprising to hear that she was a cheerleader and a member of the queen's court back at Chocowinity High School in North Carolina.

In fact, even though she claims to be a complete homebody, Allison is a go-getter like her husband. "I'm always competitive," Allison notes, without apology. Allison's college education was interrupted by a near-fatal encounter with toxic shock syndrome, which she now suspects might have had something to do with her developing MS. Soon after she recovered, she was one of 3 chosen from 700 to become a flight attendant for TWA. When she met Bruce, she gave up flying because she "didn't want to live out of a suitcase and be married." Then Bruce got a job in Dayton, Ohio, and she ran the Lancôme counter at a local department store, and proceeded to outsell the other Lancôme counters. "I had the largest counter in the Dayton, Ohio, area, so that was pretty exciting."

Now Allison has made her MS the center of her life. She has become a spokeswoman for the MS Society and the organizer of "Alli's Gators," the team from Guardian that participates in the annual MS walk. It was Allison who thought up the name, as well as the design for the T-shirts: a green-and-orange reptile wearing sneakers and a baseball cap. Last year, she fielded the largest corporate team.

When Rae-Grant first approached her about being in the Myloral study, Allison refused, fearing it might affect her ability to bear children. "I really didn't feel like being a guinea pig," she says. "He didn't push me, he didn't twist my arm." But over time she decided she "might as well do this." "If God wants me to have a child," she says, "I'll have one."

She went to see Nancy Eckert. "They took eight tons of blood and they had you consent to an HIV test. That is absolutely frightening—I don't care how good you've been. I was a nervous wreck." Then there was the "mini mental exam," the consent forms to go over, and the diary. "I probably keep the most complete diary of the study," she says. "They laugh at me." Every morning she takes her four capsules out of the freezer and swallows them on an empty stomach.

Allison Hay has no way of knowing whether she's on placebo or the real drug. But whichever it is, she is convinced that she has improved since she entered the study. Before then, she was so exhausted by the end of her workdays at the Guardian that she was considering cutting back to part-time, even though it would jeopardize her benefits. Since she began taking the Myloral, she has a lot more energy. "If it's not the oral myelin, it's my body, and that's fine too," she says. The difference was so dramatic on her last trip home to North Carolina from Pennsylvania that her mother cried. "My mom has seen me go from dragging myself off the plane to seeing me bounce off the plane."

Allison gets tears in her eyes when she talks about being so far away from home. "I like living here because I'm involved in the drug study and because I know all these people at the MS Society and have got all these great people here supporting me. . . . If I didn't have MS and I didn't have all these things to do, I'd probably be miserable here. I miss my family. I think MS has sort of become my adopted family."

Baltimore, Maryland

It's seven A.M. on a weekday, and Karyn Kraft and her eleven-year-old daughter Lindsay are well into their morning routine. Lindsay is plunked down on the living room couch with her muffin, watching *Garfield,* and Karyn is in the kitchen, making her tea. Because her movements are slow, Karyn has to allow extra time for everything: getting dressed, getting down the stairs, getting breakfast. She never knows when some small mishap, caused by the numbness in her left hand, is going to complicate things: Earlier this morning, she accidentally knocked over a glass and had to sweep up the pieces. Such are the challenges of living with relapsing, remitting MS, the kind of early-stage disease that made Kraft a candidate for the Myloral trial.

Karyn and Lindsay live in a small attached house in a development outside Baltimore called Ridgley's Choice. All the townhouses in the complex are laid out alike—two or three bedrooms up, living room and kitchen down. But some facades are brick and some clapboard, which makes it easier to tell them apart. Karyn has always wanted to live in a freestanding ranch, but she's grown accustomed to her small home on Timahoe Circle. She has generous neighbors, who shovel her walk and mow her lawn and sometimes take care of Lindsay after school. "I definitely need neighbors," Karyn points out. From the street comes the rude roar of a motor. "That's Kevin's truck," Karyn notes. "Kevin needs a new muffler." The neighbors and their vehicles are known to her.

On this morning, as on every morning, Karyn gets her bottle of Myloral pills from the freezer, counts out four, and places them on a small bare space on the kitchen counter. Every surface in the kitchen seems to be piled with papers or crowded with knickknacks, and the living room is almost as cluttered, with Lindsay's bright pink plastic dollhouse, toys, and art supplies scattered around. "I'm not the best housekeeper," Karyn allows. But she takes care of the details that matter: "Keep gas in the car, keep food in the cupboard, keep the bills paid." If there's bad weather, she explains, it's hard for her to get out to her car, and her mailbox is at the end of the road. Besides, she can never be sure when an attack will come. "You don't know from day to day. I mean I'm fine now, how will I be tonight?"

Karyn zeros in on her gold chain on the kitchen counter. "This didn't make it upstairs," she notes, as she fishes it out and slips it over her head. A wispy blonde with striking green eyes, Karyn is wearing a blue dress, belted at her small waist, with a graceful flared skirt. Between sips of tea, she talks herself through her routine. There are the cats, who need to be fed and separated—one to the basement, one into the downstairs half bath. Then she needs to write a note to the neighbor who's taking in Lindsay after school. All of this has to be done before the shorts start on *Garfield,* because that's the signal that it's time to put on coats and get out the door to drive Lindsay to the school bus stop.

Karyn Kraft's version of her life is a series of stories, told always with self-mocking humor that seems designed to ward off pity. And yet, in her retelling, there are glimpses of how difficult small things can be. There was the time she tried to give Lindsay her cough medicine, for instance. She had left the medicine half-open, since weakness in her left hand makes it difficult for her to pry open the childproof caps. And so, wouldn't you know, when she grabbed the bottle, she tripped over the kitty dishes, having forgotten there were two new kittens in the house. Then she started shaking—another occasional symptom of her MS—spilled the medicine all over the floor, and had to order a new bottle from the drugstore

Lindsay has some of her mother's gift for self-satirization. She has drawn a large picture of herself during her recent bout with in fectious mononucleosis: there are red spots all over her face from a reaction to antibiotics, blue circles under her eyes, and big teardrops are dripping down from the corners of her eyes. The teardrops are exaggerated in a way that makes them funny. Lindsay has managed to tell the story of her illness and make a little joke about it, with the tears.

Lindsay is finally feeling better, much to Karyn's relief. It's always a double worry when Lindsay gets sick: Karyn worries about her daughter, whom she calls "the light of my life." But she also worries about catching whatever it is herself. In the past, colds and flu have precipitated attacks that have forced her into a wheelchair.

This morning, however, Lindsay looks pert as can be as she pulls on her purple coat and hoists up her purple backpack to head off to sixth grade. It's a heavy load, especially since she's also carrying a clarinet in one hand, but she says it's not as bad as it could be. "It's a good thing I don't have my social studies, then it would be worse!" Lindsay's blonde hair is pulled back tight into a ponytail. She has a pearly smile like her mother's, only partially obscured by braces.

Lindsay helps her mom out with the things she can't do: When the zipper on Karyn's coat got stuck, for instance, Lindsay was the one who got it free. "Do you know what it's like to try to get material out of a zipper?" Lindsay asks rhetorically. Lindsay also bought her mom a cute, little girl's satin pocketbook with a blue plastic handle. "Lindsay gave me that," her mom explains, "because she said I was carrying too much." Now Karyn slings the long satin strap of the tiny pocketbook over her head and carries it to work every day.

Turned sideways, so that she can hang onto the railing with both hands, Karyn carefully descends the four front steps of her house and heads for her driveway. Except on icy days, she doesn't like to use a cane. Lindsay follows close behind, and hops into the car for the short drive to the bus stop. There are a lot of young families in Ridgley's Choice, so there are fifteen children waiting at the stop. Karyn is the only mother, there by Lindsay's request. "She wants me to stay," Karyn says, "so we stay."

Karyn worries about what will happen if her MS gets worse, because she doesn't want to become dependent on Lindsay. She's pleased when it works the other way: Recently she made a colonial costume for Lindsay to wear when she performs with a sign language group she's part of. Even though her fingers were too numb to feel the straight pins, she felt great satisfaction "that I was able to do it." Karyn's gone along on trips with the sign language group when they've been invited to sign in unison for various celebrations up and down the East Coast. Always, when the weather is warm and there's a lot of walking involved, Karyn knows she won't be able to manage, even with a cane. So she figures out locomotion: a three-wheeler at Disneyland, a wheelchair in Atlanta. Now that she's in the Myloral study, she also has to requisition a freezer for her pills whenever she travels.

Karyn, of course, has a story about this: In Disneyland, she got a glimpse of Pluto slipping cold packs under his fur coat to keep cool in the Florida heat. So, knowing that the heat would be intense at the Olympics in Atlanta, she got the idea of making belts with pockets for cold packs that the kids in the sign language group could wear under their heavy costumes of many nations. Karyn kept the cold packs in her freezer, next to her Myloral pills.

With barely a look back at her mom, Lindsay climbs into the school bus. Now Karyn can get back in the car and head to work, leaving the tranquillity of Ridgley's Choice for the garish strip of fast-food joints along Belair Road, en route to Northrop Grumman Laboratories on Aviation Road, near the Baltimore airport.

Karyn grew up in Baltimore—"Bawlmur" as she pronounces it in native fashion—and wanted for a while to become an air traffic controller. Her mother, she says, discouraged her because she didn't want her to travel through the Baltimore Harbor Tunnel to her job every day. Now she works as a computer programmer instead, but she still has to drive through the tunnel. And instead of the large catastrophe her mother worried about, she has daily struggles. Not long ago Karyn had a period when her left arm became so numb and weak she could barely move it. She had to push it out the window with her right arm to hand the attendant her tunnel ticket.

Like others with MS, Karyn looks back now and recognizes that there were signs of her disease long before it was diagnosed: incidents with her vision, extreme reactions to heat, episodes in meetings where she suddenly fell down. Then seven years ago, "This crazy thing starts happening at work. The left arm's going, the left leg's going, I can't move. And the guys I work with are trying to get me through it, saying, 'Are you having a stroke? Are you having a heart attack?'" That night, the doctor told her for the first time that he thought she had multiple sclerosis. The next day, the diagnosis was confirmed by an MRI.

Karyn's marriage had ended the year before she got the diagnosis. She and her husband began dating in junior high school and married young. "I realized that this other person had come into his life and it wasn't going to work out," Karyn explains. "So we separated."

Fortunately she and her ex-husband are still "good friends." The other person in his life is a man, and he has a son about Lindsay's age who is "like a brother." Lindsay sees her father on weekends, and Karyn relies on him to help out when she has a crisis with her MS. Once, when she was having an attack, he came by with soup and found her on the floor: She'd tried to get to the door and fallen down, giving herself a black eye in the process.

After about a half hour, Karyn pulls her dark green station wagon into a parking spot near the door at Grumman, a low-lying building that could serve as army barracks. She pushes the car door open, then lifts her left leg out with both hands. The right leg follows without help, and she makes her way to the entrance, passes through security, and heads down the long corridor to her office. There is stiffness in her knees and her toes turn in, so she moves a little like a wind-up toy. "Life in the slow lane," is the way she describes her labored progress.

The modular offices along the hall look about as permanent as a stage set, and many desks are empty. Karyn acknowledges that there have been a lot of layoffs recently, with the decrease in defense contracts. But her job seems secure for now. She has worked for this company (formerly Westinghouse, now Grumman) for thirteen years.

Karyn settles into her swivel chair and turns on her computer. Her chair has a large FRAGILE sign stuck on the back. But in her role at work, she mostly takes care of everybody else, helping offices all over the Northeast sort out problems with shipping and billing. It's nearing the end of the month now, a time when people at Grumman go "bonkers" because they've got so many shipments to get out. So she gets a lot of calls from people who are frustrated about tiny problems like fitting a five-line address into a four-line box and larger ones like losing a shipment altogether. Karyn calls them all "dear," and reassures: "I'm with you dear, I'm with you," she says, and promises to "play fairy godmother." She does a lot of what she calls "detective work," looking "behind the scenes" of shipping documents and finding out what went wrong.

Karyn takes care of the "coffee club" at work too, collecting everybody's dollar each week, and is the watchdog on recycling. "They're not allowed to throw out their cans," she says, loudly enough so her office mates can hear and snicker. There's a lot of joking between Karyn and the two guys she shares an office with, and it includes gentle joking about Karyn's MS. When Karyn explains that the numbness in her arm came on at the beginning of the grocery store strike and went away on the day the strike ended, Scott pipes up. "Hey! You could use that to make predictions next time." Scott is a handsome blond in his thirties who has a lush nine-hole golf course as a screensaver. Because she has double peripheral vision in her eye, Karyn sees "two Scotts" when she looks to her left. "Then I close my right eye," she says, "because one Scott's enough."

For whatever reasons—because she's vocal, because she's likable, because she's lucky—Karyn has a pretty good job situation. She's allowed to work from home one day a week, and has been able to get a computer with a special configuration that is easier for her to manage. Sometimes she has trouble double-clicking her mouse, but there's a way around that. "It's not," as Karyn likes to put it, "a show-stopper." Because she has poor circulation in her legs, the powers-that-be have even installed a special heater in the wall under her desk. "The guys use it sometimes when I'm not here," she says, "don't you guys?" The guys are momentarily quiet. "Do we look like weenies?" asks one of them. The heater has a push button so it's easy to turn on with one foot. "I can just sit here until I retire," Karyn says. In the silence that follows, everyone thinks of her uncertain future.

Karyn Kraft thinks she's been doing better since she entered the Myloral trial. But her study doctor, neurologist Hillel Panitch, isn't so sure. When she had the episode of numbness in her arm, he said he didn't think the Myloral was helping. Of course, he had no way of knowing whether she was even on Myloral at the time. But Hillel Panitch has participated in a lot of trials. That, plus a natural tendency to skepticism, makes him a lot less sanguine than the Lehigh Valley group.

Panitch, a slight, soft-spoken man in his sixties with pale blue eyes and a sad, gnomelike face, has been involved in the search for a treatment for MS for a long time. It was he who demonstrated that giving MS patients gamma interferon made them worse. From this information came the understanding that there are a group of "bad cytokines" that cause inflammation in MS, and another group of cytokines that can have the opposite effect. Panitch was also one of the first, along with University of Maryland colleague Ken Johnson, to try treating MS with beta interferon, a drug that had previously been used against viruses. When Chiron initiated trials of the interferon drug Betaseron, the University of Maryland was one of the main planning centers. By the time the drug received FDA approval, Panitch had patients who had been taking it for ten years.

When the planning began for the Myloral trial, everyone, including Howard Weiner, wanted to make sure that they avoided any suggestion that he was influencing the outcome. "Many neurologists out there think Howard doesn't control his trials well," Panitch explains, "probably because of that first cyclophosphamide study." Besides, "Howard wanted it to sink or swim on its own merits." After some debate, it was decided that the Brigham in Boston shouldn't be a site for the Myloral trial at all. Instead, there would be two key investigators, one for the United States and one for Canada. Neurologist Gordon Francis was chosen to head the Canadian group, and Hillel Panitch was chosen for the United States.

"Howard Weiner," Panitch notes, "is an enthusiast. His thing is the greatest since Swiss cheese. Other people feel that maybe there are other kinds of cheese out there." If you were searching for the perfect foil for Howard Weiner's enthusiasm, you couldn't do better than Hillel Panitch. He doesn't excite easily. After all, he points out, "we probably know 200 ways to save rats from EAE." Oral tolerance, as far as he is concerned, is just one more way: whether the same approach will work in humans is still an open question. "It's very exciting," he says, "if it works."

CHAPTER 10

The Next Question

IN THE EARLY YEARS OF AUTOIMMUNE, INC., Howard Weiner met with two or three employees at a time, helping them to understand the scientific premises on which the company was based. But by 1996, the little company had grown from a handful of brave souls to a payroll of seventy. Even though many of the employees had been involved in biotech before, few had any experience with oral tolerance. So Weiner was invited out to AutoImmune in Lexington to give a formal presentation.

It was, by now, such a familiar story to Weiner that he could tell it without thinking. At AutoImmune that day he repeated many of the well-worn phrases and slides from previous talks. There was the reminder that "the largest immune system in our body is actually in the gut." And there was the story about the very first time the postdoc fed myelin to the white Lewis rats, accompanied by the slide picturing the lively rats in the left plastic bin and the paralyzed rats in the right plastic bin. That slide, Weiner noted in his talk, had been used so many times it was cracked.

But if the talk was old, the feeling Weiner had as he spoke to the AutoImmune employees was new. All of the people in the room, he realized, were there because of his research. They were his allies in a way that no other audience had ever been. They were sophisticated enough that he didn't need to dumb down the science the way he had with the investors. But they were also, unlike his scientific colleagues, disposed to believe in his work rather than pick it apart.

He found himself relaxing and opening up with the AutoImmune employees in a new way. He told them, "You're better at running clinical trials than we are at the Center." He talked to them about a patient he had first seen when she was nineteen years old, who had since married, borne three children, divorced, and remarried the same man. When she first came to him, she had no disability. Now, he told them, "this very vivacious woman can barely walk with a cane."

Wearing a designer tie and black sport jacket instead of his usual blue blazer, Weiner rocked back on one foot and forward on the other in front of his slides, encouraging his audience to participate as he guided them through the research process. "What do you think we did when we saw this?" he asked them. And "What's the next thing you would do?" Unlike previous audiences, this one truly needed to understand the science to succeed. So when he got to the "most important experiment," the one that illustrated bystander suppression, he issued a pedagogical threat: "When I'm done explaining, I'll ask if anybody doesn't understand. And if no one raises their hand, I'm going to ask someone about it."

After the last slide, Weiner told the AutoImmune employees a story. It had to do with the day he and Bob Bishop learned that Schering-Plough was withdrawing its support. The two of them were playing golf at the Sandy Burr course west of Boston and commiserating with each other about the disappointing news. On the fifth tee, they encountered another golfer who noticed their long faces and asked what was the matter. Weiner and Bishop explained the situation. The fellow golfer was incredulous. "Are you kidding?" he said. "Do you know how lucky you are to be working on something like that? Just to be given a chance to play in such an important game is all one could want in life."

"We're all involved in a very historical endeavor," Weiner told his audience. "And I personally thank the Lord for having this happen to me in my life."

Whenever he was disappointed, Howard Weiner reminded himself that he was lucky to be alive, and to be doing what he did. It made him feel better. But in the last two years his mantra had acquired another layer of meaning. In April 1994, soon after the phase 3 Myloral trial began, the thing he had dreaded most happened within his immediate family. His seventy-eight-year-old father died of prostate cancer. The loss had shaken him, and heightened his already acute awareness of life's fragility.

Weiner got the news that his father had cancer in September 1993, as he was returning from a European celebration of his twenty-fifth wedding anniversary with Mira and the boys. "It finally happened to one of my parents," he wrote in his journal. On his visits to Denver, Howard played nurse, sleeping on the floor next to his father's bed, cleaning up after him and giving him painkillers through the night. The family relied on "Howie" always, and for the most part he liked being relied on. Now he acted as the intermediary with the doctors, passing on information to his mother and sister. The doctors told him, and he told them, that his father had two years to live.

Seven months later Paul Weiner died. Howard had flown back from Israel in time to be with his father for his last two days. At the funeral services, he spoke with his head down, knowing that he would cry if he looked at the familiar faces in the audience. The burial, in pouring rain at the oldest Jewish cemetery in Denver, seemed unreal.

Over the week that followed, memories came flooding back. One day, Howard and his sister Rhoda dropped their mother off at the Alliance synagogue, then drove around the old neighborhood, remembering everyone they knew there. Their little house on Wolff Street was just two blocks from their grandmother's, and not much farther from Aunt Gertie's. There were, besides, members of the Alliance congregation all around them. "I could walk into any house in my neighborhood," Rhoda remembers. On holidays, the streets were full of people, walking to and from services.

That Saturday, the family said Kaddish at the Alliance, which had been so central to the Weiner family's life in Denver. Howard remembered Ping-Pong competitions there before Hebrew classes, and B'nai B'rith meetings where he polished his debating skills. Once, he made it to the international competition in Pennsylvania, and had fifteen minutes to prepare a speech on the topic, "Should Christ be considered a Christian or a Jew?" He had his bar mitzvah at the Alliance, of course, and he and Mira had been married there, like his parents before him and his sister, who was five years younger, after him. Now the Alliance was about to move, following the Jews who had once populated the neighborhood around it. A new synagogue was opening in the more affluent neighborhood to the south. The death of his father and the final demise of the neighborhood were happening at the same time.

One of the things Howard Weiner's father had always said is "you can't complain if you're not in it." It was a message his son had taken to heart. Howard Weiner had been in there pitching in every aspect of his life. He had been an involved father, despite his travels, driving his sons Dan and Ron to elementary school most days and later, staying up late with them for talks about life and to help with homework. He had taken great pride in their accomplishments: he celebrated, first with Dan then with Ron, when they got into Harvard and he watched with pleasure and concern as they searched for the right relationship and made their way in their chosen careers. Dan, after earning an MBA from Wharton, was thriving in the business world and Ron was writing for television in Hollywood.

Despite a lot of rejections, Weiner had never given up on his dream of becoming a writer. After his first novel *The Children's Ward* was published in 1980, he had written three more novels, and kept trying to get them into print. Recently, he'd been pleased to find an agent who was interested and was working now on a book about MS.

Weiner was passionate about sports, as well. He was an excellent golfer, whose scores ranged in the mid-eighties, and a superb skier, although his nephew Jared claimed he could outdo him on the slopes. When he took up a new sport, like squash, he worked at it until he was competitive on a high level. Even when he wasn't a great player, he played to the death. The number of injuries he'd sustained over the years playing pickup sports was remarkable: a broken finger in a softball game, a torn rotator cuff that forced him to play squash left-handed, sprained ankles and broken heel bones, a knee injury playing touch football. This was a man who didn't hold back.

Above all, he had been "in it" in his research. "If you look at the history of the field for the last twenty years," says Stephen Hauser, the former postdoc who now is head of MS research at the University of California in San Francisco, "I don't know who has asked more right questions than Howard. And oral tolerance is vintage Howard. It's brilliant, it's creative, and it's pursued with dogged determination, despite initial laughs."

The paradox of Weiner's situation, now that phase 3 Myloral trials were under way, was that he was so "in it," as the scientific founder of AutoImmune, that he had to be out of it. To avoid any appearance of bias, none of the MS patients he and his colleagues cared for at the Brigham were to be included in the trial. And Weiner only received little whiffs of information about how the trial was going from chance encounters with principal investigators, who themselves didn't really know one way or the other.

Since the trials were off limits, there was nothing to do but pursue the research. The way to do that, as Weiner was fond of saying in his lectures, was to "ask the next question." The next question, clearly, was "How does bystander suppression work?" Ariel Miller's research had shown that there was a factor released in the gut of mice following the feeding of particular proteins, and that that factor, once switched on, was able to zero in on inflammation involving proteins in other parts of the body. Miller had taken the work a step further. He had discovered that the suppressor factor released in the gut, when extracted and placed in a dish, seemed to be a known substance called transforming growth factor beta, or TGF-beta. TGF-beta got its name from its role in stimulating growth. But what makes immunology so complex is the fact that the same factor that stimulates growth, given one set of signals, can suppress activity when it's given other signals. TGF-beta, in oral tolerance, seemed to play a suppressive role. Miller was able to show that an anti-TGF-beta agent reversed suppression in his animals, whereas other reagents had no effect.

The possibility that the T cells activated in the oral tolerance experiment were producing suppression by secreting TGF-beta was new and intriguing. But biologists weren't going to believe the finding was "real" until Weiner's lab had been able to clone the activated T cell and use it to transfer suppression to other animals. This task, it turned out, would require both skill and dexterity. That was where Youhai Chen entered the picture.

"Some people propel work forward," Howard Weiner notes. "You get a guy like Youhai Chen, he moves everybody along."

Youhai Chen was born in China but received his scientific training in Canada, where he had worked under Alex Sehon, the "father of immunosuppression."

"I'd been cloning cells for years," Yohai explains, "very tough cells that other people cannot clone. That was my expertise." When he first read a paper from Weiner's lab about oral tolerance in autoimmune disease, he thought that the idea was "too easy, too simple." Certainly it was a lot simpler than his mentor's painstaking approach. But he decided to get in touch with Weiner anyway and offer his services. In October 1992, he left Manitoba and traveled south to join the team at the Center for Neurological Diseases.

Within three months of his arrival Chen had cloned the T cell that activated TGF-beta. After two months of repetition to make sure the clone was what he thought it was, a paper about it was submitted to *Science,* with Chen as first author, Weiner as last author. It was the first of two classic papers Chen produced, with the critical contribution of Jun-ichi Inobe, during his stay in the Weiner lab. Out of Chen's work came the Weiner lab's daring suggestion that they had discovered a new class of T cells, which they called Th3 cells. Th3 cells, they claimed, were a regulatory cell activated via the gut and secreting a different cytokine than Th1 or Th2. The Th3 designation has since been taken up by some others in immunology.

Between August 1994 and July 1995, at the same time that the pivotal Myloral trials were going on, Chen and others published the article in *Science* and one in *Nature,* adding credibility to the original observations of oral tolerance. Not only was Chen able to clone the suppressive T cell, but he was also able to defuse the controversy that had risen between Weiner and Caroline Whitacre, who had been the first to observe oral tolerance in EAE. Where Weiner was able to produce tolerance by feeding, Whitacre found that she got anergy, a scientific term for cell inactivation. Chen definitively showed that the difference in findings had to do with the size of the dose that was fed, building on work done earlier in the Weiner lab by Ronnie Friedman. Low doses activated suppression, whereas high doses simply shut down target cells in the gut.

By the end of 1996, Weiner's lab had accumulated an impressive record: three papers in *Science,* one in *Nature,* and a total of thirty-five articles related to oral tolerance. Under the circumstances, it was sometimes maddening to Weiner that oral tolerance still got so little respect from his colleagues, especially the ones closest to home. He might be taken seriously at an international immunology conference, but around Harvard the subject and Weiner himself were often viewed with suspicion.

Publicity in the media didn't help. A clip on NBC's *Nightly News* about oral tolerance featured Weiner briefly, then showed two rheumatoid arthritis patients who were doing well on oral collagen. But the story made frequent references to the "chicken soup" cure, and placed Weiner at the wrong hospital besides. The idea of oral tolerance was very attractive to the general public: For several days after the story ran, the center had three secretaries working full-time answering hundreds of calls from interested rheumatoid arthritis sufferers and their families. But it was frustrating to have the research presented in such a simple-minded way. "One publishes two classic papers in *Nature* and *Science*," Weiner lamented in his journal, "and one must still deal with 'chicken soup.'"

Some in the medical community were downright hostile. When Weiner gave a grand rounds at the Brigham in the spring of 1995, a large audience turned out. In the question-and-answer period afterward, one questioner raised the old issue of cyclophosphamide and pointed out that others didn't use it because of its toxicity. A prominent Harvard neurologist asked if the slide, showing the two groups of Lewis rats, had been doctored. "He insinuates fraud," Weiner fumed in his journal. "Mean-spirited, petty, self-serving academic behavior. It's hard enough to deal with these diseases, without having to fight this as well."

Often, it was easier to deal with the company than it was with the academy. At least at AutoImmune, everyone was pulling toward the same goal. And for the most part, the company was succeeding. After a lot of back and forth, Eli Lilly had agreed to form a partnership with AutoImmune for a trial of oral tolerance in diabetes. Within a year, Lilly initiated the first of several phase 2 clinical trials to test the efficacy of oral tolerance as a treatment for humans. The trial was a one-year double-blind study involving 300 patients. Lilly also provided the oral product, called AI 401, for an NIH trial of 490 patients, which would test whether AI 401 could delay or prevent the clinical onset of juvenile diabetes.

In July 1995, the results of a phase 2 Colloral trial were positive, though not stunning. Thirty-nine percent of patients receiving a 20-microgram dose of Colloral improved significantly, compared with 19 percent on placebo. The lead statistician, analyzing the results, told Weiner and Bishop, "You won, but it wasn't a home run." Later Bishop and Weiner decided it was a "standup double."

After the Colloral trial, Bishop moved quickly to initiate a secondary offering, and was able to raise an additional $54 million for the company.

Not all the news was good. The NIH trial in uveitis, the inflammation of the eye, was disappointing, and it was decided that the next uveitis trial would have to use a recombinant human protein that would take a lot of time to produce. Also, the ABC drugs (Avonex, Betaseron, and Copaxone) would all be on the market before Myloral. This fact, however, didn't worry AutoImmune executives or Wall Street very much. The ABC drugs were all taken by injection. An oral product, if it worked, wouldn't have any trouble competing.

The wide and frequent fluctuations of AutoImmune's stock price on the NASDAQ were often irrational, in any case. The stock went as high as $18 at one point, and dropped to half that at other times. At the time of the phase 2 arthritis trial in the summer of 1995, it traded up to $14 on enormous volume. Then there was a small debacle—one that Jo Ann Wallace, who had joined AutoImmune ten months before as vice president for corporate affairs, will never forget.

Jo Ann Wallace describes herself as "a pharmaceutical junkie" who has worked in some area of the business since she graduated from the University of Texas with training in microbiology. Her first jobs were with the big boys: She traveled to South America and Ireland for Baxter, then took a job at Searle, which afforded her an office with a view of Skokie and a luxurious travel budget. Then she took a chance and moved to the more spare offices of a biotech start-up called Greenwich Pharmaceuticals, which tried and failed to bring a drug to market.

When the AutoImmune job came up, early in 1994, she was out of work, playing golf and traveling around the country visiting friends, and not at all sure she wanted to cast her lot with such a dubious venture. "I thought the concept was hokey," she remembers. Her partner Bobby Owen, however, was already working in biotech in the Boston area, and she was intrigued by the possibility that she might eventually be in charge of bringing a drug to market.

When Bob Bishop met Jo Ann, he liked her well enough that he wanted to introduce her to others. "How much time do you have?" he asked. "Well, I don't have a tee time until one," she replied. That was the right answer. After a month in Europe, Jo Ann Wallace joined the company around Labor Day 1994.

Wallace, auburn-haired and in her forties, is always perfectly groomed, even on days when she's in the office and expecting no visitors. In her work, Jo Ann Wallace prides herself on her ability to anticipate. She keeps an eye on other drug companies, watching the progress of drugs that may compete with AutoImmune's. She tries to find out ahead of time if an article is coming out in the *New England Journal of Medicine* that takes a swipe at AutoImmune's drug, so the company can come back with a defense from experts. She also thinks about how to lay legal claim to new breakthroughs in research related to AutoImmune's products.

Wallace thought she was covering all these bases in the spring of 1995, when she engineered the announcement of phase 2 Colloral findings at a meeting of the American College of Rheumatology in San Francisco. Martha Barnett, a rheumatologist involved in the study, presented the results and did "an excellent job" in Jo Ann's view. "She let us help her with it, and it really looked solid—I mean," she insists, "I spent hours on this presentation, even had

Howard look at it. Everyone felt comfortable with what we were saying, that we weren't underselling, we weren't overselling." Even though academics tend to be skeptical, she was pleased that "there wasn't any hostility in the room." She was also pleased that the stock analysts she knew about were planning to file basically positive reports. She went to bed feeling confident.

At dawn she received an alarming phone call. The investment firm of Dillon, Read was getting ready to file a report that Martha Barnett's presentation was "all puffery" and that AutoImmune had oversold the results. Soon after, Wallace got hold of the report. "Colloral Data Presented at Amer. College of Rheumatology Disappointing," the headline read, "Rating Lowered to Neutral." The Dillon, Read report, although acknowledging that the results looked positive using one mode of statistical analysis, suggested that they faded to insignificance using other statistical methods.

It was an arcane argument that made Jo Ann suspect evil intent. There is money to be made in short trading, borrowing shares from a brokerage firm and selling them with the expectation that they will go down in value and you can buy them later at a lower price. There were a lot of short positions in AutoImmune stocks at the time. This is often true of biotech stocks, because they are highly risky and as likely to go down as up. It is thus difficult to prove that such scheming is going on. Besides, Jo Ann had a self-professed tendency to suspect the worst.

Whatever the merits of the Dillon, Read analysis, the "morning note" that went out over the wires on that October Friday quickly drove the stock down six points, from $14 to $8. Wallace, who had become "physically ill" when she read the report, got on the phone to Bob Bishop, back in Boston, and urged him to call other biotech investment banking companies and ask them to "do something, say something, to support the stock."

Over the next few days, several complied. Hambrecht and Quist issued a "spot report" five days later, concluding that "overall we feel the data is excellent—and that the bear arguments surrounding the data are unfounded." UBS Securities weighed in the day after, noting that the stock "has taken a sizable downturn for a series of spurious reasons." "We have analyzed the new information," wrote UBS broker Tim Wilson, "and see nothing in it to change our positive opinion of the product."

Because of the Dillon, Read incident, Jo Ann Wallace has vowed she will never come to a meeting without knowing which investment brokers are in attendance who might be likely to report on AutoImmune. "I want to know who's there, what they look like, what their name is, and I want them to know I'm available night and day to talk to them." Some time later, when Wallace got a call from the young woman at Dillon, Read who filed the report, she unleashed her fury. "What do you look like?" she demanded. "FAX me a picture of yourself! I don't ever want to be in a room with you without knowing it!" Not surprisingly, the Dillon, Read analyst didn't comply.

The stock recovered, slowly. Then it was hit again, this time from the academic side. In early December 1996, Weiner was awakened at dawn by a phone call from England, asking him for a comment on a *Science* article that was appearing later that day. Issues had been raised, it seemed, about the safety of oral tolerance. Five Australians, including pioneering immunologist Jacques F.A.P. Miller, had engineered a diabetic mouse which, when treated with an oral tolerance regimen, seemed to become sicker. Within two weeks a second article appeared in *Science,* this one with Weiner's colleague and friend Stephen Hauser as senior author, reporting that injections of MOG (one of the myelin proteins) could result in an exacerbation of an MS-like disease in the very small marmoset monkey. In the same issue, an editorial warned that "oral administration of antigen, shown to be successful in mice and now being tried in clinical trials, may under some conditions enhance disease."

Weiner responded by pointing out the artificial nature of the Australians' mouse model and the many differences between the marmoset experiments and the human trials then in progress in MS. "We have closely monitored all the patients in human clinical trials of oral tolerance therapy (now more than 1,400 subjects) and to date have not seen any indication of adverse effects on the disease process or systemic toxicity," he wrote in response to a colleague's query. In BioCentury, a biotech investment report, Weiner emphasized that there was no evidence of a buildup of antibodies like those described in the *Science* articles in human patients. Although his rejoinders were vigorous, stocks are often sold on whiffs of doubt. AutoImmune stock was down by late March to $11.

In the end, as everyone at the company acknowledged, the stock price would be based on very little until there were phase 3 results. If the Myloral trial were successful, the stock had the potential to skyrocket. Everything depended on these results, which would be announced sometime in the spring of 1997.

During the last weeks before the announcement, Weiner spent a lot of time hoping for the best and preparing for the worst. He told himself and everyone around him that no matter what happened, the work on oral tolerance and MS would go on. "There's no question in my mind, it has to work," he insisted. "It may not work this time, it will work the next time. So in some ways I'm not afraid."

But Howard Weiner's gut contradicted his words. "I have the same feeling in the pit of my stomach," he confessed, "that I had around the time that my father died. There's this thing in your stomach. And you're kind of waiting for it to go away. It's very momentous, there's no question about that."

Montreal Neurological Institute

By all accounts, neurosurgeon Wilder Penfield was a man of many parts, a French horn player and an expert at billiards. But his greatest passion was reserved for the human brain. And so, when he managed to accumulate the funds from Rockefeller and elsewhere to build a neurological institute in Montreal in the 1930s, he devoted the entryway to his first love. A visitor, entering the main doors of the Scotch baronial–style edifice, is drawn into a waiting room that uses the brain as a decorating motif. The ceiling is a web of neuroglia cells, spidery shapes first drawn by the great Italian neurologist Golgi. "The cells themselves . . . stand out black," Penfield noted with pride, "while tiny blood vessels and coloured background complete the picture." Every inch of the reception hall is brain-inspired. The border around the ceiling is a pattern drawn from the cerebral ventricles, and the iron gratings over the radiators are based on a drawing of nerve fiber by a French neuroanatomist. The furnishings are neurological as well: There are standing lamps with segmented necks that resemble the vertebrae of the spine and a center table with a wood inlay depicting a cross-section of the brain's two hemispheres. If the visitor sits on one of the art deco benches and stares at the floor for a while, he will realize that it too is part of the scheme: Its pink-and-black marble inlay is a geometric interpretation of the spinal cord.

In Penfield's day, the real interior of the brain was on view upstairs on the fifth floor, in the state-of-the-art operating suite. There he conducted operations that attempted to map what he called the "undiscovered country." Under local anesthetic, Penfield performed surgery on epileptic patients, laying open a portion of their brain. "The operative wound," he explained, "was carefully ringed about . . . so that the patient's face and body were fully exposed below the operative field. Thus he could be observed by those who sat beyond the sterile barrier and could converse with them." As research fellows watched from the amphitheater, their faces pressed against the sloping plate glass, Penfield used electric stimuli to activate various responses. In a procedure that was not painful to the patient, Penfield discovered that if he stimulated one area, the patient made a sound, whether he wanted to or not. Stimulus to another area activated the little finger, another the left side of the lower lip. Perhaps most surprising, Penfield was able to stimulate certain areas that caused several patients to reexperience a vivid memory from the past.

In 1997, the Montreal Neurological Institute that Penfield founded, known throughout Canada as the MNI, was still renowned for its probing of the brain. But the high drama of those earlier explorations in the operating room is no more. The knife has been replaced by powerful magnetic forces that affect protons in the body, and powerful computing forces translate the flux of protons into a picture of the soft tissue of the brain and spinal cord. The surgeon's queries and the wide eyes of the research fellows have been replaced by the penetrating electromagnets in the tunnel where the patient lies, very still, as technicians activate MRI, magnetic resonance imaging. Later, the neurologist will look at a film that shows him what the machines and computers "saw."

The institute was the gathering place for the MRIs of every patient in the Myloral trial. All the trial sites around the United States and in Canada sent their films—snapshots in time of each patient at the beginning of the trial, at one year, and at the end of the trial—to Montreal. And when the last patient's last MRI had been taken and sent off to Montreal, the Myloral trial would officially end.

Entirely by chance, the last patient to enroll in the Myloral trial and the last patient to finish was a twenty-nine-year-old graduate student named Nicholas H. who lived in Montreal. So the last of the several thousand sets of MRI films taken in the Myloral study didn't have to be packed up and sent by air. Instead, it was hand-carried down the hall.

Nicholas H. grew up in Greece and arrived in the northerly climate of Montreal at the age of seventeen. He sometimes wondered if he might have avoided MS altogether if he hadn't left home. A marathoner who had his first attack after a long run, he wondered too if the running might not precipitate attacks. Or if the infectious mononucleosis he contracted shortly after arriving in Canada might have something to do with it. Most of all, he wondered about his future.

Nicholas was a tall young man with a handsome aquiline profile, a radiant smile, and large brown eyes with long lashes that had a slightly stunned look in them. Even though he came from the Mediterranean, he welcomed the cold of the Montreal winter, and was well prepared for it, moving quickly along ice-layered sidewalks in his hooded jacket. "This is my ideal time," he explained. "All my attacks have been in the summer."

The summer after he entered the Myloral study, Nicholas didn't have an attack. That made him hopeful, and he took up running again. But then he had two attacks in fairly quick succession that caused numbness in his legs. After that, he hoped that he was on placebo. Like everyone else in the trial, he wanted to believe that Myloral could change his life.

Nicholas's graduate work was in experimental psychology, so he studied the brain and was able to follow developments in MS research. But that didn't make it any easier to know what the future would bring. Because of his MS he felt more pressure to finish his Ph.D. and get established somewhere. He wanted to remain in academia, because it was what he loved and because he believed the academic community would be tolerant of someone with a disability. Yet he sometimes thought it would be wiser to get a job in the pharmaceutical industry, because that way he could make more money quickly and save up in case he became disabled. "I mainly get sensory symptoms," Nicholas explained. "But what if the next symptom was a motor symptom, and I couldn't work very well? You always think about that."

On Tuesday, March 18, 1997, at a little before two in the afternoon, Nicholas H. walked up the hill to the MNI, the limestone fortress Penfield built, and headed for the MRI unit. There he climbed up onto a narrow concave table and lay perfectly still as a technician slid a white plastic basket, resembling a birdcage, over his head. Then the table slid slowly into a tunnel so narrow that he could feel the sides of it along his arms and hips. There was a microphone inside the space, so patients who became claustrophobic could ask for help. But Nicholas, the calm scientist, managed his session in the dimly lit MRI tunnel without too much difficulty. He lay immobilized in the tubular prison, and waited for the noises of the MRI, created by the magnet's pull on the machine's metal parts, to begin. First came a tock, tock, the sound of a ball-peen hammer tapping somewhere near his head. Then a louder, steadier sound like a pneumatic drill doing double time. Then silence, the ball-peen hammer again, and other sounds, this time more like the alarms that plastic security tags set off at the door of a department store. Then doubling of that sound, so it began to resemble a Phillip Glass piece. After forty-five minutes, the set of films of Nicholas's brain was complete, and the Myloral trial was officially over. At long last, the researchers conducting the trial could take off their blinders. And all those who had placed their hopes in Myloral could find out if it worked.

CHAPTER 11

Results

WEEKS BEFORE THE ANNOUNCEMENT of the Myloral trial results, Bob Bishop had begun to spread the word among contacts on "the Street" that there was going to be an event in late April. "The rules of the road in dealing with the stock market," says Bishop, "really say don't surprise people. And if we surprised people by doing it [the announcement] two months earlier than they expected, they feel we haven't been candid with them, and they can't trust us." On the other hand, it was risky to give an exact date for the announcement because "if we don't get it done by then, they assume there is a problem." So even though he was fairly sure the announcement would come on Monday, April 21, Bishop hedged. The word went out that it could be that Monday or it could be the following Monday.

Saturday was the day chosen to unblind the trial, because it gave the company plenty of time to prepare a press release and a strategy before making an announcement on Monday morning, as the market reopened. "Myloral is considered our lead product," Bishop explained, "the one that is closest to market approval, closest to being able to give us positive cash flow from operations. And so a positive or negative result from our phase 3 trials is a very material event from a stockmarket perspective."

The final piece of planning Bishop did was recreational. The results were to be announced in Durham, North Carolina, at the offices of the Cato Institute, the consulting company that had coordinated the trial. Knowing that no one would be able to concentrate on Friday, the day before the announcement, Bishop got the idea that it would be "fun" to play golf at Pinehurst, a premier golf course not far from Durham. Golf was, Bishop admitted, about the only activity that took his mind off AutoImmune. Besides, he reasoned that "no matter what happened there was going to be at least one wonderful day that weekend."

During the second week of April 1997, Weiner and the executives at AutoImmune got the word: The Cato consultants had finished analyzing the data and were ready to "break the code." On Saturday, April 19, they would gather at Cato to learn the results of the Myloral trial. That Friday, Howard Weiner got up at 5 A.M. to take an early flight. He left a message for Mira on the erasable white board on the kitchen fridge. "I'm off to meet one of my fates. Must have some positive findings and a new era will be opened. You're with me in my heart. Love, Howie."

At about the same time, Bishop and AutoImmune financial officer Mike Rogers, a young banker who had replaced the original CFO, Tom Hennessey, were making a presentation at an investors' conference in London entitled "Planning for Success." The conference was sponsored by Union Bank of Switzerland (UBS), a major investor in biotech. UBS biotech analyst Tim Wilson had been a consistent fan of AutoImmune, which he saw as an "excellent buying opportunity." The next day, Bishop and Rogers flew across the Atlantic to North Carolina and the moment that would prove Wilson right or wrong.

On Friday morning, a nervous and excited quartet of AutoImmune executives teed off at the Pinehurst course. It was a perfect golf day, and the foursome—Weiner, Wallace, Bishop, and

Rogers—played thirty-six holes. Nobody played particularly well. "I think we were all a little shaky," Bishop admits. Weiner was off his game. Mike Rogers, the only one wearing his AutoImmune sweatshirt, had the best score, which was taken for a hopeful sign.

That night the two "key investigators," neurologists Gordon Francis from Montreal and Hillel Panitch from Baltimore, arrived at the hotel. They had flown in from Boston, where they'd been among 6,000 people attending the forty-ninth annual meeting of the American Academy of Neurology. The meeting was called "Revolution in Neurology," because of the advent of new treatments for neurological disorders. But the talk at the meeting, at least among individuals working in multiple sclerosis, had been about disappointing drug failures. There had been a negative result in the trial of a drug called sulfasalazine and a less than thrilling result with a drug called cladribine. Most dramatic was the failure of a $40 million study of Linomide, underwritten by Pharmacia-Upjohn. There were actually three multicenter trials of Linomide under way in the United States, Canada, Europe, and Australia that involved 1,681 MS patients. But the trials were stopped after side effects caused two deaths and eight heart attacks. "Everyone is hoping for a positive Myloral trial," Weiner wrote in his journal. "The neurologic community needs a lift."

Saturday morning, Howard Weiner was up and writing in his journal a half hour before the hotel wake-up call. "The trial was always so much in the future," he wrote, "1997 seemed so far off. . . . Then it slowly moves forward. Then it is next spring, next month, next week, tomorrow, and now we are only thirty minutes away from driving over to Cato to see results. As I write this, thinking of only thirty minutes, I get chills that run through my body and the queasiness returns. . . . It will be an enormous relief to have the data in front of us. No more thinking about how to deal with the unknown, there will be something to put one's hands on and we can move forward one way or the other."

Weiner wrote on and on about the developments in the research, the possible outcomes and the consequences, his excitement building all the while. "What greater thrill than to be facing this moment and having the chance to help people with MS? Sitting and talking to all the MS patients for the past twenty-five years, now there is the chance that I will have done something that will truly benefit them. Not everyone gets that opportunity. My prediction: positive results that will end in a product, or very strong trends that will allow things to go forward, though it may require another five or even ten years of work and new trials. I am ready and not afraid." Then he invoked the Hebrew prayer: "We thank the Lord Almighty for keeping us alive, sustaining us and bringing us to this day."

The meeting at Cato was called for 9 A.M. At 8:45, Weiner was in the lobby of the hotel with his video camera, greeting the team from AutoImmune as they gathered to take the van to the offices of the consulting company. "It is showtime!" he announced into the camera mike. Bob Bishop appeared. His usually red face had taken on a little extra color from the golf course.

"Who's that, who's that?" Weiner was like a dad filming his children. "When do we get the results," he asked as Bishop rounded the corner, "is it next year we get the results?"

Bishop smiled indulgently at Weiner's lame joke. "The way I look at it, it's about fifteen minutes."

The group climbed into a rented van for the short drive to Cato. Bob Bishop, CEO, was at the wheel. The clock on the dashboard read 8:56. Weiner was wired. He talked about whatever came into his head—the flowers, the name of the road, the best iron to use for a golf shot taken from the van to the front door of the Cato Institute, the brick building that now came into view.

"What would you use from here, Mike?" he asked. "A seven-iron?"

"I'd use a p equals .041 iron," Mike Rogers answered, alluding to the statistically significant p value they were all hoping to see in the results.

"Seconds away," Weiner said as they pull up to Cato. "Nanoseconds," Jo Ann countered.

And Weiner, "We're ready to rock and roll."

Then they were in the building, following arrows that led to the conference room, where a long oval table awaited them. Seconds after they walked in, Malcolm Fletcher appeared. He was grinning so widely that Bishop and Weiner whispered to each other that the news must be good. But he looked even more pale than usual, and his "I'll just sit down here," had a sort of forced gaiety about it.

Fletcher suggested that they go around the room and introduce themselves.

Weiner started it off, "Hi, I'm Howard Weiner, how are you all?"

People tittered, since Howard Weiner is the one person that everyone already knew.

Bishop made a sort of awkward little joke: "By the way, this is the guy who started this stuff."

Under his breath, Malcolm Fletcher could be heard whispering, "I don't want to delay this thing too much longer." And then the room fell silent, and Malcolm Fletcher had the floor.

"The first thing I have to say, ladies and gentlemen, is I'm afraid the news is not good." Fletcher's voice sounded breathless, as though someone had just punched him in the stomach. "I'm sorry to . . . I'm sorry to point that out." Fletcher added the usual qualifications—the data was preliminary and some of the numbers could shift by as much as 10 percent. But, lest anyone should cling to hope, "the conclusions we're presenting are accurate."

It was the final stage of a four-day ordeal for Malcolm Fletcher. He had left for the Cato Institute the previous Wednesday, three days in advance of the announcement of trial results. It was his job, with the help of statisticians, to take the data that had been organized and summarized by Cato, and translate it into terms nonstatisticians could comprehend. He knew from experience that this would be a complicated task. "With clinical trials," he explains, "it's not like flipping a coin or getting the numbers out of a lottery container. . . . Usually you get some clear signals and you get some mud. And you have to poke around a little in the mud and make sure that the conclusions that you're pointed toward are sound. That usually can take some time—I mean days or weeks."

Because the results were critical to the fate of AutoImmune and to its price on the stock market, Fletcher had to work more quickly than he would have in a large pharmaceutical company that was less reliant on a single product. He also had to work in secret, to avoid leaks and rumors. So long before the day he left for Durham, he worked out a plan that would make speed and stealth possible.

In the first place, he decided to bring along an independent statistician to help him make sense of the data. Cato, the consulting company, would come with their statisticians, of course, but Fletcher wanted his own expert to supplement his rudimentary math skills. "I know my limitations," he explains. "Not bad at looking at the shapes of data. But the actual arithmetic basis of that I'm not strong at." He invited Phil Lavin, a Harvard School of Public Health specialist in biostatistics who had his own statistics company in Framingham, Massachusetts, to fly down with him and be his "right hand," going over the results along with the people from Cato.

To do this without being noticed by the press or interested parties, Fletcher, who feared his name might be recognized in the small world of the Durham Research Triangle, asked a colleague to book rooms for him. As a further precaution, he chose an out-of-the-way motel called Duke Towers, where "you'd never find a pharmaceutical executive." Duke Towers had been a Liggett tobacco factory, and its rectangular shape adapted well to its second life as a motel. For $60 a night you could rent a suite, complete with kitchenette and sitting room. It was, according to Fletcher, a well-run place, with nice plantings and late sixties motel decor, including drapes that weren't too bad "if you sort of half-closed your eyes."

Fletcher and Lavin settled into Duke Towers Wednesday morning. The three representatives of Cato came to them soon after their arrival, bearing perhaps fifteen pounds of paper. Fletcher assumed the Cato team already knew the trial results. Yet for some reason, they didn't seem eager to talk about it. "They chitted and chatted, and they put down their piles of paper. And then they started off trying to say a saga."

Fletcher finally cut them off. "What's the bottom line?"

The answer was surprisingly unqualified. "The news," Cato statistician Imogene McCandless told him, "is not good."

Malcolm Fletcher liked to talk about the moment in 1992 that he became a believer in oral tolerance and the hairs stood up on his neck because he realized "there's biology there." Now the Cato statisticians were telling him that the "biology" wasn't there after all, at least not in this trial. Fletcher had also imagined a "lifetime competence" that would end his money worries. But "in a few milliseconds it was clear that wasn't going to be." For the rest of the afternoon, he was too stunned to concentrate on the data.

That evening, he went out to one of his favorite restaurants in the research triangle with Phil Lavin and a physician friend. "I had a couple of really good belts of Pernod, and suddenly things looked a whole lot brighter." Later that night, realizing that he had been too "shell-shocked" to absorb any of the details of the data earlier in the day, he understood his mission.

When the officers of AutoImmune and the key investigators gathered in the conference room at Cato, he needed to give them the news quickly and clearly, so that no one could postpone a reckoning with the painful facts. "We didn't have the luxury. Everybody was going to experience the loss. But they needed to go to the bottom line immediately, so that they could do what they had to do that day." Fletcher had brought along portable printing equipment so that he wouldn't have to go out to a copy place and risk being spotted. Now, with some of the enthusiasm he had expected to devote to positive results, he began to prepare a handful of overheads that would tell the disappointing story.

When Saturday morning finally came and everyone was gathered in the conference room, Fletcher moved quickly from his announcement that "the news is not good" to a short series of slides that provided the big picture.

The first slide compared the attack rate of patients on Myloral with those on placebo: There was virtually no difference. He showed the result first in numbers and then as a graph. Then he broke down the results for each of the four subgroups, including the males in the immune group who had responded so well in the phase 2 trial. Once again, the results were the same for each group, and none were better than placebo. Then he showed the astonishing slide—the one that had people shaking their heads in disbelief for weeks to come: both the patients on placebo and those on Myloral had a greater than 56 percent reduction in their attack rate!

This was a puzzlement and in one way a slight embarrassment for the ABC companies. There was a general feeling in the MS community that hope is an important factor for patients, and probably has biological consequences. But in other trials, improvement with placebo was never higher than around 35 percent. In those trials, the improvement in patients on active drug was in the 50 percent range, around the same range as the placebo group in the Myloral trial. In fact, the placebo worked as well in this trial as two out of three active drugs currently on the market.

Why was the placebo rate of improvement so much higher with Myloral? There were several possible answers. One had to do with the waxing and waning of the disease. Patients selected for the trial were required to have had two attacks in the previous two years, which may have made them likely to have fewer in the subsequent two years. But the more convincing explanation had to do with side effects. In previous trials with Avonex, Betaseron, and Copaxone, it had been fairly easy for patients to tell the difference between placebo and drug. All those drugs were injected, and patients on active drug often had reactions at the injection site. They sometimes had flulike symptoms as well. Patients talk to each other, and they tend to figure such things out. It seemed likely that those on the active drug knew it, and were more likely to expect improvement than those on placebo and therefore to do better. With Myloral, however, there were no side effects and there was no way of knowing the difference, so everybody was equally hopeful. Also, patients *wanted* to believe in Myloral. It was safe and "natural," and there was no needle involved.

But a 56 percent improvement on placebo? These were, as Bob Bishop was to tell the press on subsequent days, "unusual" results. They even caused one investor to ask Bishop if he were sure the subjects in the trial really had MS! The answer to that was a definite yes, of course. But the results did seem to suggest that optimism and frequent monitoring by good physicians and nurses was an enormous factor in the disease. All in all, since Myloral had no side effects of any consequence, it could be said that most of the 500-plus MS patients benefited from being in the trial. This, however, provided small consolation to the devastated group gathered around the conference table at the Cato Institute.

Fletcher was as succinct as he had hoped to be. His presentation took no more than half an hour, and by the end of it "they had enough information, if they were composed enough, to function." But it would take several hours to get to that stage.

Mike Rogers, who had been videotaping at Howard's request when Fletcher started to speak, turned the machine off soon after he heard the words "the news is not good."

Around lunchtime, Howard Weiner picked up the video camera. "We're gonna film the bad with the good," Weiner announced, "we're not afraid of it because we're going to succeed in the end anyway."

"There's nothing we can do about it," Weiner observed, "we've got to go by the biology. And we go by the Winston Churchill quote, 'never ever ever ever quit.'"

The Churchill quote was one of the ones on Weiner's office wall, where it was described as "The Shortest Commencement Speech Ever Given." Over the next several days, it became a mantra. Bob Bishop joined Howard in chanting it now for the record—"never ever ever ever quit."

"Gotta keep going," Bishop said. He was sitting at a computer drafting a press release, but he was blue around the mouth.

When the camera was trained on him, Weiner looked nearly as shattered. In the morning, he had worn the Kelly green shirt he'd gotten when he went to the Masters golf tournament. It was supposed to bring good luck. Now he'd put a pin-stripe shirt over it. All the exuberance of early morning had been knocked out of him. He sat hunched over the table with a hand on his forehead, as though he didn't want to let people see his eyes. His assessment of the results was brutal. "It became very clear that there was just no biology there, I mean it was like the feeding of water . . . why didn't it work? The only thing I can think of has to do with dose. And we have to go back and look at dosing. It's not going to be easy."

Then, with smiling resignation, "I'm going to spend the next couple years, talking at conferences and trying to explain why it didn't work."

Weeks later, when the results of the MRI data came in, it was possible to actually discern some "biology" in a particular group of patients. But on this Saturday, everyone was impressed by how "flat" the data was, how little evidence there was of any effect. Malcolm Fletcher noted that the consistency of the data, the lack of complications or contradictions, was owing in part to the fact that the study was well designed and well executed, so that the results were consistent across centers. "In a way we would have been better off if there had been a little more noise. We could have spun things our way."

Before he flew down to Durham to learn the results, Fletcher had slipped a $100 bottle of champagne into the fridge at his condo. It wasn't the kind of indulgence he permitted himself often, since his alimony payments left him without a lot of spare cash. Nonetheless, a good result in the Myloral trial would have been grounds for a celebration.

But when he arrived home on Sunday morning, Fletcher was "needing hugs," not champagne. To make the painful task of calling all the principal investigators bearable, he set himself up on his deck. "It was a beautiful day. I actually went and got the warm umbrella, got cushions for the deck chairs, got myself real comfortable. One of the few times when one's wife is actually prepared to give you everything you want—she brought me cups of tea and meals. And then I just relentlessly went down the list. And what I did with them all was I very quickly told them it was bad news. I did a description of the results and then I said that we would need to shut the thing down. And I tried to allow for the fact that they were going be shocked."

Howard Weiner arrived home later that day and videotaped a last scene in his kitchen with Mira. "Back in the womb," he commented from behind the camera. "There's Mirele." He aimed the camera at the optimistic note he'd written on the refrigerator board before he left predicting "a new era."

"Well, it didn't happen," he noted with a little laugh. "But let's see where we go next."

On the morning of Monday, April 21, Patriots' Day, the sky was cloudless and bright blue. The temperature was cool enough to please the runners assembling in Hopkinton for the annual Boston Marathon and yet not too cool for the families gathering on the Lexington green to celebrate the battle that began the Revolutionary War. Many people had the day off. A few miles from the festivities on the green, however, at AutoImmune, Inc. no one was in a holiday mood.

Bob Bishop was behind his big desk, in his customary telephone position, leaning way back in his chair, phone in left hand, right hand behind his head, one leg crossed over the other at right angles. The press release announcing the trial results had gone out over the wire at 7:30 A.M., and the impact, when the stock market opened at 9, was expected to be dramatic. Bishop was talking on the phone to as many of the company's major stockholders as possible to give them a personal heads-up before the opening.

"There was an incredibly high placebo rate," Bishop was saying. "To put this in perspective for you, the placebo rate was as high as two of the three FDA-approved products."

Bishop was talking to Jackie Doeler from the investment board of the University of Wisconsin. She had invested in 1.5 million shares for the state and stood to lose millions of dollars in about half an hour, when the market opened. He was trying to reassure her that they were not lost forever: There would now be an "orderly shutdown" of everything related to Myloral so that the company could focus on Colloral, the rheumatoid arthritis drug. "We're set up in a way that we get two bites out of the apple."

Though his face looked pained, Bishop's voice was deep and forceful, as it needed to be. All day long, he would offer stockholders variations on these same themes. "We're disappointed, but not dissuaded" became a favorite, as did "we're obviously very chagrined." And "it's a failure of the product but not of the hypothesis.

Oral tolerance is alive and well." And, "we'll be stepping up to the plate again real soon."

Bishop had moved on to a second stockholder conversation when Howard Weiner arrived carrying his beat-up canvas briefcase. Weiner's face was red, whether from sunburn or emotion, and his smile tight. The stockholder Bishop was talking to turned out to be Kip Agar, a Midland oil executive whose daughter has MS. "Let me talk to Kip," Weiner said. He walked over to the speaker phone and added, "This was one experiment and it didn't work. We're already thinking of the next experiment." And a little later, "We're gonna figure this out. Mother Nature is a tough cookie, but she'll eventually yield her secrets."

It was going to be a day of high hurdles, especially for Weiner and Bishop. At nine they would meet with all the employees to tell them the bad news. Then at ten there would be a telephone conference with investors. At one o'clock, they would meet with the company directors to discuss whom to lay off and how to go about doing it.

Fred Bader walked into Bishop's office to ask what he could do to help. Soon after, Jo Ann Wallace and Malcolm Fletcher appeared. Bader and Fletcher recited Churchill again: "never ever ever ever quit." Weiner was starting to talk about the next experiment. Anthony Slavin had been getting good results using myelin in combination with other cytokines back in the lab. Maybe they could get some IL 4 and use it as a "synergist" with Myloral in a new trial? Fletcher looked at him in disbelief. "It would be hard to get funding for that right now," he said gently, "because of perception."

At minutes to nine, Weiner, Fletcher, and the rest of the officers headed down the hall and up the stairs to the meeting with employees. The meeting room was no longer big enough to contain the AutoImmune payroll, which had expanded in recent months to over ninety. Several dozen employees stood at the back of the room, spilling out of the doors into the hallway.

Few knew what to expect, and the room was instantly quiet when Bob Bishop rose to speak. "This is a difficult day," he began. Then he read the press release.

> AutoImmune Inc. today announced that preliminary analysis of its Phase III trial for Myloral showed no difference in response between active treatment and placebo. The Company will conduct additional analysis of the Myloral data before deciding what, if any, further efforts will be taken with this product. AutoImmune currently intends to direct its resources toward the development of Colloral, its product for rheumatoid arthritis, which is nearing completion of three Phase II clinical trials. Preliminary results of these trials, which involve more than 800 patients, are expected to be announced next month.

"We are very, *very* disappointed," Bishop said, stating the obvious. Then he went over some of the same ground he'd covered with investors, but with additions, for this more knowledgeable crowd. "The trial was exceptionally well run. We got very tight results, with comparable data from all the centers." He noted that MS requires very long trials, and for this reason "we made a strategic decision not to do a dosing trial." He acknowledged the company might have done differently if they'd had the benefit of a "retrospectoscope," an invented instrument that was to become everyone's favorite in the coming weeks.

Bishop explained that the company would start an "orderly close-down" of Myloral-related work and focus all energy on the phase 2 trials of Colloral. When he mentioned that certain projects would be "mothballed," everyone in the room knew what that meant. The first question, when he finished his remarks, was the one on everyone's mind.

A man in his thirties, seated in the front row, asked, "Will there be layoffs?"

Bishop's usually robust voice was shaky and he seemed close to tears. "I think it's going to be unavoidable. Nothing has been decided yet. We will be developing a severance package policy."

Later Bishop would cite this as his most excruciating moment among many. "I've had to do difficult things before, but this, to stand in front of the employees—I had the opportunity to come here as the first full-time employee, so I regard everybody who's here as if they're part of my family. And from that perspective this was more difficult than anything I've done before."

Howard Weiner walked up to stand by Bishop and provide moral support. When he gave his formal talk to the employees, he had dressed up. But on this day he was in his usual well-worn blue blazer. "We took a hit," he says. "It's clear that oral tolerance is very real. But we have to go back and understand why this didn't work. I feel the worst for the people who have the illness. We're ultimately going to succeed, I'm convinced of that."

Fred Bader, who has a knack for finding a reassuring anecdote, offered the story of Avonex, one of the MS drugs now on the market. The giant phamaceutical Schering-Plough had no success at all with it and gave it to Biogen thinking it was worthless. Biogen managed to demonstrate its effectiveness for MS and bring it successfully to market. "It's not uncommon in the pharmaceutical business to have these cycles," says Bader. "The work you've all done won't go to waste."

There were more questions and discussion. Someone wanted to know why the federal safety review board, which takes periodic looks, didn't stop the trial on grounds of "futility."

Bishop explained that the trial "didn't hit the criteria. We hung in there every time."

He ended the conference with the pronouncement that "oral tolerance is real" and thanked everyone for "all the work you've done so far."

After the meeting, Ahmad Al-Sabbagh, the Lebanese-born scientist who prepared the bovine brain used in the first trials nine years earlier, approached Weiner. Ahmad was now working for AutoImmune, Inc., where he had more responsibility and better pay than in his early days in the Weiner lab. Now he told Weiner, "I have a broken heart."

Weiner replied, "we all have a broken heart."

Back down in the executive offices, Bishop was still recovering from the employee meeting as he prepared to host a conference call with investors. "I can be less emotional with the investors, I don't know them and love them." His face was still red and there was a crack in his voice.

"Now how do I do this?" he asked Michelle Linn irritably. Linn, a consultant from the public relations firm Feinstein Kean Partners, was in charge of logistics for the investor conference call. Bishop was told to dial a number. He tried it and an impersonal voice came back through the speaker phone, "Good morning. And welcome to the AutoImmune preconference call. Please identify yourself."

"I'm Bob Bishop, the CEO of AutoImmune."

In the next room, chief financial officer Mike Rogers was at his desk, listening in on the conference call and watching the stock market online at the same time. On top of his monitor there were two small wire figures on a seesaw. When one went up, the other went down. On the screen, Rogers was keeping an eye on AutoImmune and its competitors. Since the market opened, AutoImmune had dropped nine points, from 13 3/4 to 4 5/8. Rogers switched to Biogen, the maker of the MS drug Avonex. Biogen was up two points. This was Mike Rogers's baptism by fire, his first biotech crisis after years in the quieter world of banking. His eyes were wide behind his round rimless glasses.

Bishop began the conference call by reading the press release to investors. He told them the study was well run, and recited the now familiar phrases—a failure of the product, not the hypothesis, and so on.

"The coin didn't fall our way," Tim Wilson of UBS noted.

Another investor asked, naively, if AutoImmune couldn't go ahead and apply for product approval despite the placebo effect, since there was a 56 percent reduction in attack rate.

Yet another, who represented the large investment held by the Tisch family and clearly wanted to protect it, asked Bishop about other plans. This gave Bishop a chance to accentuate the positive: He mentioned not only the Colloral trial, but also the partnership with Eli Lilly in diabetes and the beginning of enrollment in a trial to test oral tolerance as a method for preventing rejection in human organ transplants. "We're not out of it yet," Bishop told them.

There were seventy-four people on the line for the conference call, which lasted thirteen minutes.

The night before he learned the results, Howard Weiner dreamed that he was in a house full of people, including his sister, and had to make a call to find out about the Myloral trial. The results, he learned from the phone call, were negative, but he resisted telling anyone. He kept phoning back, "to find out about nuances, but every time I am told that it is just negative." In the dream he then went upstairs in the house and encountered another large crowd. The images were so vivid that, for a while after he woke up, Weiner believed the dream was true. Even after he realized it was just a dream, it brought home to him "how devastating it will be if the trial is totally negative or only shows a little bit."

Now the nightmare had become reality, and Weiner needed to tell the crowds of people. He had already tracked down David Hafler in Atlanta the night before. Today he would have to work his way down his list of people who had given support, encouragement, and, in many cases, years of work to the dream of an oral tolerance treatment for MS. It wasn't going to be easy.

Since he had no office at AutoImmune, Weiner commandeered the conference room. He sat alone with the phone at one end of the long gray table circled by empty chairs, pulled out a rumpled piece of yellow lined paper, and began dialing the names on the penciled list.

One of the scientists on his list was Caroline Whitacre, the biologist from Ohio State who first got oral tolerance to work with

EAE in rats. Whitacre had sometimes resented Weiner, and she worried that he was rushing too quickly into human trials and might jeopardize her research as a result. But when Weiner reached her, she greeted him with genuine pleasure.

"Hi, Howard. Happy Patriots' Day!"

When he told her the news, she said, "oh my gosh." She worried, she told him, that "the disbelievers in oral tolerance will point their fingers and say I told you so."

"I'm not afraid of them," Weiner answered.

Weiner defended, as he would over and over from now on, the decision to go to phase 3 trials so quickly, without a dosing trial.

Whitacre was gracious enough not to say I told you so. "You have to ride the wave," she said. "Investors have a fairly short attention span."

One of the hardest calls Weiner had to make was to his mother. Since he last saw her, she had become extremely ill. An enlarged tumor in her lung required chemotherapy, and sent her into despair. For a while, she was saying she wanted to die, and could receive the news of the Myloral trial "upstairs." With time, however, she recaptured some of her old spunk, only to be brought low by a cardiac problem, possibly a complication of her cancer treatment, which had landed her in the hospital. When Weiner talked to her the night before, he decided she was too sick to hear the bad news. But now he called again, determined to tell her.

"Hi honey," his mother said. Then she started right in explaining what had happened with her heart. Her doctor was in the room, and she wanted to pass the phone to him so the two experts could discuss her case.

"OK, but before you do that, we have the results on the MS trials." He told her the disappointing news.

"Oh no," she responded.

"Science is tough," he told her.

"You didn't lose that much, did you?" she asked. "You protected yourself, didn't you, Howie?"

Howard's mother was referring to his decision, made early on, to sell half of his AutoImmune stock. He had decided that he would benefit from the patents, if the trial succeeded, and needed to limit his financial risk.

But of course, there was no way Weiner could protect himself from this loss, which was far more immense than any monetary investment could ever be. He talked to the doctor about his mother's heart condition.

"Don't be too disappointed," his mother told her son, as they said good-bye.

"I'm all right, don't worry about me," he answered defensively, as a son does when he doesn't want his mother feeling sorry for him.

Weiner tried next to get LA information so he could reach Samia Khoury, a neurologist and researcher at the CND who did some of the earliest work on oral tolerance and EAE. But instead of dialing 213 he dialed 215. "I'm off by one variable," he pointed out. "That's probably what's wrong with the Myloral trial. We're off by one variable."

While he was making his calls, Ahmad Al-Sabbagh entered the room apologetically. "I wanted to ask you something, but I don't want to take your time," he said. He was agitated. "I wanted to ask you am I going to be laid off?"

Even though he worked at AutoImmune, Ahmad still believed that Weiner was his ultimate boss.

"It's not my decision," Weiner pointed out. "But I can give you a 95 percent assurance you're not going to be laid off." Ahmad had tears in his eyes as he thanked him. "I appreciate it. I hope your faith in me is justified."

After he left, Weiner looked crestfallen. "That's what everybody's worried about—are they going to be laid off? He has four kids, he's worried. The investors lose money. The Tisch family loses millions. But they have more millions. With employees it's their jobs, and with patients it's their lives."

On the bulletin board out in the hallway, there was a remnant of the optimistic days before the trial, when AutoImmune was gearing up for commercialization: an advertisement from the April 6 *Boston Globe,* in color, offering a large number of new positions. But a few feet away, in a corner with a picture window and tables where people collect and chat, the talk was all about finding the next job somewhere else.

"Hey, what's happening?" said the first thirty-something.

"Not much, just updating my CV," said the second.

The first had been checking for jobs on the Internet. "I've got leads on positions in Boulder and in the Bay Area. I don't want to go back to Boulder so I hit the reset button right away. The Bay Area will be a hard sell at home—we just moved from there seven months ago."

As they spoke, Bishop and his key executives were meeting in the conference room, figuring out which employeees to lay off to reduce the workforce at AutoImmune by 40 percent.

For a long time, ever since it had seemed likely that the code would be broken on that April weekend, Howard Weiner had known that he would have the results by the time of the first Passover seder on Monday night, April 21. As usual, he and Mira would be celebrating at the home of one of his oldest friends, a child of refugees from Germany who had grown up with him in Denver. Weiner had imagined arriving at the seder that night with good news. But it was not to be.

The Weiners had a reputation among their friends for being prompt, but they were the last to arrive that night. The seven other guests had learned already of the results. All of them had in fact invested in AutoImmune, and taken together had lost thousands of dollars that day. But no one, seeing Howard Weiner's face as he arrived, dared to broach the subject of their losses, his were so obviously greater. His gaze was vacant, preoccupied, and he barely engaged in conversation, except when it concerned AutoImmune. Nor did he trouble to follow the seder ritual about when to eat what, instead breaking off little pieces of matzoh and nibbling distractedly as the premeal Haggadah passages moved from one reader to the next. His mood was catching. After the seder dinner, everyone agreed to dispense with most of the postmeal ceremony. The door was opened briefly for Elijah, and the three most familiar songs were sung. But as the friends gathered up their dishes and divided the leftovers, the subject returned to AutoImmune and the trial results. Weiner stood in the front hall, repeating the words that seemed to help the most: "It's one experiment."

The next morning the *Boston Globe* described the results as "a serious setback" that "leaves AutoImmune years away from profits," but added that, according to analysts, it is "too early to write the company's obituary" since they are still "flush with cash." The investment firm Hambrecht and Quist described the results as a "serious blow to the company," but added, "while many investors are interpreting the Myloral failure as an indictment of the entire oral tolerance approach, we remain cautiously optimistic about the future of Colloral based on positive results from previous trials."

Montgomery Securities, which had earlier characterized AutoImmune as "one of the true home-run hits in the biotech group," now downgraded its rating from buy to hold. Its biotech expert David Crossen predicted that Myloral was "most likely dead" and that the company "will reduce its spending rate drastically in order to stave off having to shutter."

The most dramatic market reaction, and some of the sharpest criticism, occurred in Britain. Reuters reported that "fall-out from the failure of a new U.S. biotech drug hit Britain's fast-growing biotechnology sector on Tuesday," with the "most heat" on companies about to publish "important late-stage drug data, notably

Celltech Group and Cortecs International." The failure, Reuters noted, "follows a string of high profile disappointments in the US" and underlines the risk of investment in biotech, a "confidence-based sector" in which "confidence has taken a bit of a tumble." Reuters quoted an analyst who said that AutoImmune had "pushed its product faster than was warranted." The outcome, all analysts agreed, was "significantly positive" for Biogen, producer of the MS drug Avonex.

Surprisingly, even though there had been a 56 percent improvement on placebo, no one in authority anticipated the possibility that the negative results could have a "reverse placebo" effect on patients in the trial. One board member had raised the question on the day of the announcement in North Carolina, and others had mentioned it in passing. But it did not seem to occur to anyone that some thought should be given to how patients were told the news.

Hillel Panitch, when asked if he had been given guidance, responded that "I don't really think that we need very much, because the study is not being stopped for any safety reasons." But Nancy Eckert, in Allentown, wished for more advice. "We've had no instruction on what to tell people," she complained. "I had one woman come in who is just on thread's edge, I look forward to seeing her every month to check up on her. And today I didn't want to let her go, she was so upset. She said, 'Now what do I do?'"

In the absence of advice, Nancy Eckert relied on her own good instincts. "I've told them think about all the attention you've paid to your body in the last three years. You've taken care of yourself in so many ways." And then, because "you can't just send them away empty-handed," she gave everyone a packet with information about "diet and spasticity and other things" along with her card and the assurance that they could call her anytime.

Allison Hay got the news that the trial had failed on Tuesday. Her first responses were characteristically upbeat: "Oh well. It was a good thing that we did, right? Well, guess what, now my husband has no reason to say you can't get pregnant until this trial's over with. But then again, it may just scare him that much more that treatment's not going to be available." Later in the conversation, she said, "When I sit down here and start thinking about it, who knows? I may start crying."

A week later, she went in for a termination visit with Dr. Rae-Grant. "I turned in my three-month supply of Myloral. We discussed what the future holds, planned a visit to his office in six months, and then I was on my way back to work. It was really sort of sad. I won't see Nancy on a regular basis anymore. I actually looked forward to my visits. I've actually been fighting some pretty strong fatigue this past week. I hope this is just a temporary thing." A few days later, she reported that her energy was back, following a "huge adrenaline rush" from talking about "living with MS" on a local radio talk show. "I think my fatigue was due to being disappointed. But I'm sure there is light at the end of the tunnel."

Karyn Kraft heard about the trial failure from Kathy Conway, the nurse practitioner for the study at the Maryland site, and she found it jarring. "My arms and legs go right to the freezer in the morning," she said. "It felt strange not to do it." She worried about having a relapse, if the drug was helping her. And when she heard about the high placebo rate, she wondered "what's going to happen now, everybody's hope just went down?"

Karyn had been doing a little less well over the winter. After a fall that left her bruised, she'd taken to using a cane more often when she was outside. The weekend after she got the news, she began to move even more slowly than usual and to have a lot of trouble with her balance. She was on "open label" by then: She had completed the study, and was taking the drug. So she called

Kathy Conway to see if she could take the Myloral she had left in her freezer. Kathy Conway said she could use up the pills she had, but wouldn't be able to get a refill. "I'm not sure whether they will help or not," Karyn said, "but I was getting very concerned that I was getting worse. Maybe it was because I had been out cutting my lawn. Maybe it's psychological too, who knows?"

In June, Karyn finally learned that she had been on Myloral all along. "I'm still not convinced that it didn't help," she wrote. "The last two weeks I've had a couple of bad falls. It sure makes me wonder since this has started about a month after I went to taking two pills a day [instead of the four she took during the study]. But who am I? The experts know best." She had decided to put off taking Avonex, though her doctor advised it. "I'm going to try something in a pill form containing algae. It has no side effects, seems to really help some people, and costs less."

In Montreal, Nicholas H.'s first reaction was "Well, that's research, but I have to tell you I feel very disappointed." A week later, he wrote that "in my mind the published results of the preliminary trial were absolutely spectacular and I thought that Myloral had the potential to surpass in effectiveness any of the current medications. I'm disappointed and it's not so easy to get rid of this feeling because to my knowledge there is no realistic hope for a new treatment in the near future. . . . I still think that oral tolerance has potential."

In mid-May, all the Montreal patients in the Myloral trial were invited to a wine and cheese gathering to hear a discussion of the results. Dr. Gordon Francis presented, and another doctor translated into French for the francophones in the study. "The whole talk was very negative about the oral tolerance approach," Nicholas reported.

Six weeks later, at the end of June, Nicholas had another MS attack while he was rushing to complete his dissertation. He decided, reluctantly, to go on Betaseron, the only drug approved for MS in Canada.

In the days immediately following the announcement, Howard Weiner refused to give an inch on the decision to do a phase 3 Myloral trial. "I would have done exactly the same thing again," he insisted, "because it was the clearest path to getting results and getting something for the MS patients. You could say that there was horse A and horse B, and I bet on horse A. If I could go back, I would bet on horse B. But that isn't how science works."

Weiner tended, during those early days, to repeat himself a lot. He would go over and over the decision, and cite the comments and names of various people who told him he had done the right thing. He also dwelt heavily on the enthusiasm people felt for oral tolerance. He insisted that the phase 3 results in Myloral, which everyone had viewed before the announcement as so critical, were nothing more than a setback along the way.

"This is not a reason to quit," he insisted. "Because the basic underlying theory hasn't changed. We're trying to climb the mountain and the car broke down, it doesn't mean you aren't going to get there. This wasn't a dichotomous event. What happened is that one of the paths you tried didn't work and there are six other paths to get there." He even suggested that it may have been better to have a failure. "Because we would have gotten something on the market and it would have worked but it wouldn't have worked great . . . we'll get to an effective drug quicker because we know so much more about how to do it."

Howard Weiner knew that it was lethal, in science, to look discouraged. If he seemed to give up, his lab, his reputation, and the oral tolerance idea would suffer as a result. But it was also true that Weiner had a hard time admitting defeat in anything he tried, least of all this, the biggest gamble of his career. Privately, far from viewing the Myloral failure as a minor speed bump, he associated it with death, particularly with the death of his father. "It's like somebody in the family died," he acknowledged. "People actually offer condolences and use the word *condolences*."

Weiner would never dwell on the idea of the failure and death for long, though. "If somebody in the family dies," he pointed out, "you can't bring them back." Oral tolerance would come back. Still he admitted to feeling apprehensive, as he thought about the upcoming results in the arthritis trials. "It will be much, much harder to explain. If the arthritis trial is negative, I will truly have to be tough to carry forward."

CHAPTER 12

Not the Last Supper

May 12, 1997, AutoImmune, Inc., Lexington, Massachusetts

In the coffee room at AutoImmune, there had long been a poster of a cartoon duck in giant sunglasses, plunked in a lawn chair and sipping his drink through a straw, only vaguely aware of the bullets whizzing over his right shoulder. In the past, the poster looked like management's little joke, aimed at loafers on the job. Now it isn't so funny. Since the Myloral announcement, the employee roster has shrunk from nearly 100 to 51.

On this Monday morning, the survivors gather in the conference room to hear the results of the trials on which the company is pinning its last hopes: the seventh and eighth of nine phase 2 Colloral trials in rheumatoid arthritis. When the results of the Myloral trials were announced, some people had to stand; now many chairs are empty. As CEO Bob Bishop works his way to the front of the room to make the announcement, everyone seems to expect the worst. "I made the first cut," a young chemist says softly as he settles into a chair near the back. "I don't think I'll make the second."

Bob Bishop, his sizable belly filling out his starched white shirt, holds a company press release in one hand and a paper coffee cup in the other. "The last time I spoke to you in this room," he begins, "I felt physically shaken. I don't feel that way today." He starts to read from the press release.

There were some who criticized Bishop and the company for their gloomy public statements immediately after the Myloral trial. One of those was the key venture capitalist and chairman of the board, Barry Weinberg, who insisted that "the releases the company put out were too self-denigrating." Wall Street, according to Weinberg, doesn't handle black or white well, they need shades of gray. And the press releases on Myloral were unnecessarily black.

Bishop probably has Weinberg's criticism in mind as he addresses the AutoImmune employees on this May morning. His assurance that he is not "physically shaken" may be part of an attempt to be more measured about these results. The results are gray, rather than black. But, on the other hand, they're a long way from white.

"AutoImmune Inc. today announced preliminary results from two Phase II trials of Colloral for rheumatoid arthritis," Bishop begins. "Both trials demonstrated positive trends for Colloral, although statistical significance was not demonstrated versus placebo." The announcement goes on to mention that there is one more phase 2 trial yet to be completed, with results expected in July.

The press release has managed to be upbeat about some unimpressive results. But when Bishop, Weiner, and the vice presidents got the news two days before, the reaction had been far less sanguine. The same group who had traveled to North Carolina for the Myloral results repeated the ritual for the Colloral results, driving out on Saturday to Boston BioStatistics in Framingham to meet with the statisticians who put it all together. Malcolm Fletcher, once again, was the bearer of disappointing tidings. And once again, they all tried to collect themselves to take the next step: telling the story to everybody else. That evening, there was a dinner at Bob Bishop's house, for Weiner and the officers to talk over the results and regroup. Bob and Susie Bishop posted a sign on their door. It read, "This is not the last supper."

Now, on Monday morning, Bishop is trying to pass on the same message to the employees. "We didn't have a clear win," he tells the assembly, "but we didn't have anything as devastating as Myloral." He points out that these phase 2 trials were the seventh and eighth in the series, and that previous trials were more positive. "We've had six successive trials which had a positive effect on this disease." There is also a ninth trial still in progress, and that may yield positive results. What's more, the company still has a lot of cash, over $40 million, in fact. It will be more than enough to carry through a phase 3 trial of Colloral with 600 rheumatoid arthritis patients. "There is potential for a good drug here," he insists.

After the meeting, Bob Bishop has to face the phones again, as he had after the Myloral trial. He recycles some of the same phrases. "I'm chagrined, but not discouraged," is one of them. Between callers, Bishop has the habit of correcting his posture, pressing both hands down on the arms of his swivel chair to raise his large bulk partway up, then easing slowly down with a straighter back. Half to himself, and half to anyone within hearing distance, he is beginning to develop a new theme that he'd hinted at during the meeting with employees. He said then that there may have been "too much of a country club approach" in the last two trials of Colloral. There may have been "too much rescue medication available, and not enough hard-nosed you're in and you're out." Now, somewhat mysteriously, he keeps repeating that "trial design is the name of the game."

Inside the company, a lot of people know what he's talking about. There have been problems with the most recent trials, problems you wouldn't want to tell "the Street" about, because they would be seen as further proof that the company is on the way out. To insiders, however, the fact that the recent trials were flawed gives hope. Maybe the results are not the fault of the drug but of the way the drug was tested.

Everyone in the company pretty much agreed that the Myloral trial had been extremely well run. Malcolm Fletcher noted, ruefully, that the lesson of Myloral might be that "you can have a group of dedicated, smart, almost brilliant people and they can get together around a promising undertaking, they can do everything absolutely right, and it can fail." The one thing wrong with the Myloral trial was the decision to go for broke, jumping to a phase 3 trial without doing the standard phase 2 trials. Of course, in retrospect this was everyone's opinion.

The company wisely balanced the big gamble in Myloral with a conservative approach to their other drug, Colloral. And they saved around $40 million to see the Colloral trials through to the end, without having to go back to the market for more. Now they were faced with a paradox: The Myloral trial was probably the wrong trial, but done right. The Colloral trials were the right trials, but some of them, particularly the trials labeled 007 and 008, were done wrong.

The 007 trial was the last in a series that honed in on finding which dose of Colloral was most effective against rheumatoid arthritis. In many trials, the big question is safety: How much of the drug can you give before the side effects outweigh the benefits? This, however, was not the question with Colloral, a uniquely safe drug. The question was efficacy. It had been shown, even in the earliest animal trials conducted by Cathy Nagler-Anderson, that too much of the drug could shut off the oral tolerance effect. And too little drug would have no effect. What was the optimal dose?

The 007 trial was a six-month trial of about 400 patients with severe rheumatoid arthritis who were tested against placebo on three different doses of Colloral: 5, 20, and 60 micrograms. In this trial, according to Malcolm Fletcher, one doctor actually "subverted the study" by "deciding it would be reasonable if the patients dropped out after three months. He said that six months is just too long." He actually "propagated this thesis to the point where an awful lot of patients did drop out." Despite the problems with 007, it provided useful information about optimal dose. The study suggested that the best dose of Colloral was 60 micrograms.

The 008 trial, as Malcolm Fletcher admitted in a candid moment, was "a mess." It was a mess for a number of reasons. For one thing, it was begun in haste, before the results of the 007 trial came in. Here, once again, money pressures affected decision making. The 008 trial used a 20-microgram dose, a best guess from earlier trials, instead of the superior 60-microgram dose. As Jo Ann Wallace acknowledges, "we jumped the gun."

The wrong dose was only one of the problems with 008, however. There were also inconsistencies in the way the doctors at the various sites around the country carried out the study.

The study was intended to compare Colloral to placebo in a trial of 200 patients with severe arthritis who had been taking Methotrexate, one of the most effective drugs on the market. The patients were asked to withdraw abruptly from Methotrexate to enter the study, which would last for six months.

The study began well enough: The patients stopped their Methotrexate and were placed randomly in Colloral and placebo groups, in double-blind fashion. But as time went on, and some patients began to have difficulty, doctors allowed them to go back on Methotrexate. That was where confusion set in. Doctors at some centers kept their patients in the study, even though they had gone back on Methotrexate, whereas others dropped patients on Methotrexate from the study. The result was "noise" that greatly reduced the power of the 200-patient study. This inconsistency was what Bob Bishop was referring to when he said there wasn't enough "hard-nosed you're in or you're out."

As he sat in his office on the day of the announcement, Bishop struggled to come up with the meaning of the results for the future of the company. On one hand, the trials showed positive trends. Also, even though it wouldn't be wise to tell Wall Street, they might have shown more if they'd been better run. Then there were the previous positive trials. In baseball terms, Bishop surmised, shoving his horn-rims up against his large, rosy face, the trials were probably a single. Or, on second thought, "perhaps we had a man on first, the batter singled, drove him to second." But this didn't quite satisfy Bishop either. "The worst face you could put on this would be that we had a man on first, there was one out, the batter hit a long fly ball, it was caught, and it moved the man to second. But now we had two out."

Wall Street decided AutoImmune had at least two out, or maybe even three. After the Myloral failure, the stock had lost 69 percent of its value. After the Colloral announcement, it lost another 42 per cent, ending the day at 2 9/16. The company was now trading in dreaded "drill bit numbers." The press, despite the upbeat company release, emanated gloom: "AutoImmune's Arthritis Drug Comes Up Short," read the *Boston Globe* headline the next day. "AutoImmune Inc. yesterday suffered its second serious setback in three weeks when it admitted that Colloral, its proposed oral drug to treat rheumatoid arthritis, was statistically no better than placebo in two phase 2 trials." David Crossen of Montgomery Securities was quoted as saying that "having two drugs not work as expected raises questions as to whether oral tolerance is yet another interesting theory that may prove to be a nice idea, but nothing else."

This was a particularly cruel blow: Montgomery Securities had been one of the investment bankers, along with Hambrecht and Quist, that organized the original Bishop and Weiner fund-raising road show for AutoImmune back in 1993. Montgomery's David Crossen, a biotech veteran, had always been one of the most positive of analysts when it came to AutoImmune. Before the Myloral trial failed, he had called AutoImmune a home run. Now he was saying oral tolerance was just a "nice idea."

The Montgomery Securities quote gave fair warning that it wasn't going to be easy for AutoImmune in the coming months. In particular, it injected an element of dread into the preparations for a big investment conference in July. The meeting, which would take place at the deluxe Pierre Hotel in New York, was being put on by Montgomery Securities. There Bob Bishop, along with the heads of most of the biotech companies in the country, was scheduled to make his case.

July 16, 1997. En route to Manhattan. Today CEO Bob Bishop needs to look successful. He's put on his dark blue suit and his custom shirt with the RCB monogram on the cuffs, and he's riding into the Pierre Hotel, scene of the Montgomery Securities Health Care Conference, in a limousine. The limo actually turns out to be less expensive and more convenient than a taxi, which makes it ideal: rich-looking but cheap. AutoImmune, Inc. is on a very tight budget.

After the announcement of the Colloral results in May, Bishop had to initiate a second wave of layoffs at the company. The employee roster has now shrunk from fifty-one to twenty. Fred Bader, the vice president for operations, has left to take a job at Centocor, a biotech company in Philadelphia that is on the way up now, after some hard times. Most offices outside the executive corner are now empty, and there is a plan to move soon into smaller quarters in the basement of the building. Meanwhile, visitors who come to the door have to be buzzed in, and everyone, including the CEO, answers his own phone.

As the limo driver navigates a complicated route to Manhattan along small Queens streets of modest brick rowhouses, Bishop opens the three-ring notebook on his lap and goes over his talk one more time. Bishop downplays the importance of the talk, which is scheduled for 4:45 that afternoon. "I don't know how many will be there today," he says. "People tend to leave early for cocktails." Yet he has put a lot of time, with Vice President Jo Ann Wallace's help, into preparation. They've restructured the talk, after the recent disappointing results, eliminating what are called "builds," a series of slides that parcel out the information to the viewer in sequence. The first slide shows baseline information, and the following slides look like the first but add more information. Builds make a nice visual effect, and Bishop likes to use them. But they also leave information up on the screen a long time. You don't want to do that if the news is bad. Just now, builds don't seem like a good idea.

Since the Colloral announcement, Bishop has had to explain himself in meeting after meeting. In general, the biotech pros have stayed calm about the situation. Eli Lilly has remained committed to the diabetes trials they're conducting in partnership with AutoImmune. The board itself, meeting two days after the Colloral announcement, reacted by planning the next step. The venture capitalists on the board had seen it all before. Barry Weinberg said, "I don't think it's as bad as everybody says it is." Alan Ferguson, who was involved early on behalf of the 3i company, saw the results in both Myloral and Colloral as reason to stay invested, despite the big stock drop.

"It's unfortunate," Ferguson noted, "that every move they make gets scrutinized with a lot more intensity than it would if it was buried in a large pharmaceutical company." Ferguson, who used to be in drug development at Johnson and Johnson, says that if he were still at J and J and looked at the Colloral results, "I'd say it's interesting. We've got enough data now to do the next phase."

The meeting Bishop and the officers worried most about was the annual stockholders' meeting on May 15, always an unpredictable event after a loss like the one AutoImmune had just absorbed. When a stock drops from $16 to $2, there are a lot of ordinary people who lose money. And you never know which stockholders may show up at the meeting. Many companies get hit with shareholder lawsuits. Vice President Jo Ann Wallace and other officers of her previous company, Greenwich, had been sued after their drug failed, and she definitely didn't want to go through that again. Even without lawsuits, which so far hadn't materialized, there were likely to be angry shareholders at the meeting asking tough questions.

The stockholders' meeting, in the downtown offices of Bank of Boston, turned out to be a pleasant surprise. The only tough question of the afternoon came from a young man with slicked-back black hair whose cell phone identified him as a professional investor. He asked, in a crisp British accent, why there was such a large disparity between the results of the phase 1/2 trial of Myloral, where there was a 30 percent reduction in attack rate in thirty patients compared to placebo, and the phase 3 trial, where placebo and drug both produced an even greater reduction in attack rate. Bishop didn't really have an answer to this question, but he speculated. Patients in the phase 1/2 trial had no expectation that the drug would work, whereas those in the phase 3 trial had high hopes, thus raising the placebo effect to equal what may have been an uneven drug effect. Not surprisingly, the questioner looked dissatisfied.

Still, on the whole, the meeting had been much easier than anyone anticipated. Bob Bishop, who had been "expecting rotten tomatoes," was grinning as he walked out. "I needed for this day to be over," he said.

As spring turned into summer, the *Boston Globe* ran an article under the headline "For biotech firms, trials and tribulations." The piece, by Ronald Rosenberg, focused on pivotal failures in the biotech industry—more than two dozen in the previous two and a half years and five in Massachusetts in the previous six months. And of all the failures, the one Rosenberg chose to emphasize was "AutoImmune Inc.'s highly anticipated oral medication to treat multiple sclerosis." On the inside page, there was a photograph of Bob Bishop in shirtsleeves and striped tie, leaning on a bench in an empty lab. An executive without employees, peering out into the future with a wan smile. By the time the *Globe* article ran, two weeks before the Montgomery conference, AutoImmune's stock had dropped to 1 5/8.

Were further proof needed of AutoImmune's current low standing, it would be the location of Bob Bishop's talk at the Montgomery Health Conference. The Wedgwood Room is the smallest of the three rooms used for the conference. It is down a winding stair and difficult to find, in what Jo Ann calls "the basement." Still, the room itself is pleasant enough. Like everything else at the Pierre, it seems overdressed, with its loges and mirrored walls. Overhead, a huge recessed chandelier is surrounded by a pastel blue ceiling decorated with white clouds.

AutoImmune vice president Jo Ann Wallace looks worried as she scans the long white tables in the half-empty room. If Bob Bishop is jocular and upbeat to a fault, Jo Ann Wallace makes up for it by expecting the worst. Surprisingly, since she is the public face of the company, she's often the one who puts things the most bluntly. When Bob Bishop decided to come to the Montgomery meeting, she imagined him "out on a plank, with everyone in the audience sharpening their saws."

Bishop is seated at a table up front, with an empty chair at his side intended for David Crossen, the Montgomery Securities broker and former ally of AutoImmune who is supposed to introduce him. A large sign behind Bishop reads, in large illuminated letters, "MONTGOMERY SECURITIES: THE POWER OF GROWTH." At the back of the room there is a digital clock, counting down Bishop's allotted time in large red numbers. Bishop waits briefly for Crossen to come, then decides to introduce himself. "I'm Bob Bishop, the chairman of AutoImmune."

Bishop is off and running. "No one," Jo Ann Wallace observes, "presents better." His fluent commentary is perfectly synchronized with his slides, and he finishes with a minute and two seconds to spare on the digital clock. The real challenge, as it turns out, comes in the "breakout session," a gathering of a few people in a smaller adjoining room who want to hear more about the company. A man in his thirties who sticks out because he's wearing an open shirt and silk jacket instead of the universal dark blue suit and tie, sits in the front row and lobs questions. "What would you say were your two biggest mistakes?" he asks.

Jo Ann Wallace, sitting in the back row, looks alarmed. But Bishop, in informal mode, clearly likes the challenge. A big man on a small chair, he leans forward with legs apart and levels his face with the questioner's. He doesn't admit to mistakes, but instead defends the decision to go into a phase 3 trial with Myloral. It would have been just as expensive to do a phase 2 trial, and in fact "it turned out to be a very big phase 2 trial."

What, the questioner wants to know, would a rheumatologist have to say about Colloral? "I think he'd probably tell you that it was somewhat helpful to many patients. But there were a few patients it *really* helped. And if it were available he would use it."

The young man in the silk jacket remains cocky, sure he can put Bishop on the defensive.

"I assume shareholders' lawsuits have hit?" he asks.

"Not a one," counters Bishop.

"So you're really in crisis, hunker-down mode," he suggests.

Bishop replies that he's done all that is necessary to bring Colloral to a phase 3 trial. "We've taken strategic action," he counters. "I don't call that hunkering down. I call that focusing."

Back in Boston the day after the conference, Jo Ann Wallace and Bob Bishop have entirely different impressions of the presentation, from start to finish. In the first place, Jo Ann Wallace is convinced that David Crossen, the Montgomery broker, stayed away on purpose because he didn't want to introduce Bob Bishop.

"He's got egg on his face," she reasons, because he backed AutoImmune and now it isn't doing well. "He did that deliberately, don't kid yourself."

Before they went, Jo Ann had worried about AutoImmune even being at the conference, and running the risk of "overselling, overpromising," and bringing on a lawsuit. "Right now, there is a lot of anxiety among our shareholders," she noted, "both retail [small, individual investors] and institutional. And it'll be some charlatan attorney that whips a retail base into a frenzy and that's how a lawsuit will get filed." She was concerned that Bishop's natural buoyancy would get him in trouble. "Bob is so enthusiastic and he's so positive about everything that even if there's just a little thread of something there when you get done listening to him you think there's a lot more."

She also suspected the young man in the silk jacket with all the challenging questions might be preparing a lawsuit. "I was very nervous, honest to God. I thought, 'he's taping this whole thing and he's getting ready to go back home and call his attorney and say OK, here it is, let's go.'" Even though Bishop was clearly triumphant in the exchange, Jo Ann was aware that this one questioner was "hogging and pissing off other people in the room." Sometimes, she notes, Bob gets on a roll with one person and doesn't notice that others aren't listening. "He needs to read body language better." All in all, Jo Ann Wallace concludes, the effect of attending the conference was "probably neutral."

Bob Bishop sees things differently. He brushes off the absence of David Crossen, who appeared later and who probably just got tied up somewhere. "He would have been there if he could," Bishop insists. "He headed a biotech company himself, so he understands." As for the questioner, Bishop took his calling card, which said he was working for a portfolio management group in Santa Barbara, California, at face value. As far as Bishop is concerned, he was just doing "due diligence" for clients who might want to invest. Bishop thinks the conference had a positive impact. "The stock," he notes with a grin and a shrug, "is up to 3 today."

At first, Howard Weiner was relieved when he heard the results of the Colloral trial. At least they weren't totally negative. But then, when it registered that they were only weakly positive, he began to get "that sinking feeling again," the one he had when his father died and after the Myloral announcement.

It would be like that for months. He would be going along fine, busy with his demanding schedule of travel and meetings, and then something would trigger new waves of feeling about the disappointing results in the trials. At the lab, he struggled to be upbeat, knowing he needed to keep up morale. "I can't show my mood to people," he explained. "That doesn't help."

He was testy at times, however, and easily provoked. When Thomas Spahn, a German postdoc working on the role of gamma delta cells in the gut, talked about "the bottom line" in a Friday lab presentation, Weiner bristled at the capitalist phrase. "We're not interested in the bottom line here," he snapped. "We're scientists, we go by the biology."

At the same lab meeting, he was impatient with Ruthie Maron when she questioned the significance of one of Spahn's findings. "There's clearly a significant result there," he declared, streaking the red dot of his laser wand back and forth across one of Spahn's graphs. Never one to be cowed, Ruthie grinned and shook her head no. Howard wanted significance somewhere, anywhere, but she wasn't going to give it to him.

Because he was trying so hard to maintain an even keel himself, it upset Weiner when others treated the trial failures as the ultimate tragedy and told him how sorry they felt for him. "I don't need that," he would say. When a postdoc described the atmosphere in the lab as "grim," he was so angered he had to leave the room to collect himself. When he came back he pointed out that two postdocs in the lab had gotten grants from the MS Society *after* the failure of the Myloral trial. The oral tolerance research was going strong.

Equally upsetting were complaints from friends who had invested heavily or even whimsically in AutoImmune and demanded to know what they should do next. Once, after a friend went on too long about his losses, Weiner resorted to the mind game he used to make himself feel better. He reminded his friend of the Holocaust, and of a mutual friend whose son had died of a brain tumor. Next to those losses, what were a few thousand dollars, or hundreds of thousands? Such comparisons quickly ended the complaining. They also made his friends realize just how catastrophic the trial failure must seem to Weiner, despite his outward display of confidence.

The hardest thing by far was seeing patients after the trial failures. "I didn't expect to feel that as much as I do," he said not long after. "It's painful anyway, to see these people. But to be there really on the battlefield and to realize that you didn't succeed in something that would have brought all this hope and all this excitement and a new infusion to these people. And I know the Myloral wouldn't have helped everybody, but if it helped a certain number of people that would have been very good."

So many of the patients he saw on Tuesday mornings had been coming to him for years and getting steadily worse—the thirty-seven-year-old who was now paralyzed and came in a wheelchair pushed by her poor aging parents, the woman who could barely sit up and who sobbed because she had just come back from Florida and feared she could never manage to get there again, the proud young man of thirty, so clearly in trouble, who insisted he was "a thousand times better," then had to push off the walls to complete his ambulation test. Over and over in his journal, Weiner wrote about how "frustrating" and how "sad" it was to sit and listen to patients' questions and stories and have nothing new to offer. He hated feeling helpless.

Yet, paradoxically, seeing his patients could sometimes help Weiner put things in perspective. One Tuesday morning, he saw a woman in her thirties who had traveled from Puerto Rico with her husband for a consultation. She had signs of early MS, but she was still walking normally. Her mother, however, also had MS and was very seriously ill—bedridden with contractures. Clearly this was a family in which there was a genetic disposition to MS.

The Puerto Rican woman had heard about oral myelin and wanted to make sure that Weiner was still working on it. "I'm all right for now," she told Weiner, "but I have teenage daughters. You have to find something for them."

That was the sort of thing that got Weiner's juices flowing again. After all, there was a whole new generation of MS patients coming along. Oral tolerance might not work now, but it would work eventually. "I think that it has to work," he would insist, addressing his listener and himself at the same time. "We know that it is not the strongest vaccine, it is a relatively weak vaccine it turns out, so you make a stronger one. And the potential is enormous. Enormous!"

One of the things that helped most was thinking differently about the time frame for success. If he used the analogy of boxing, which he sometimes employed with his patients, he could say that he had lost only this first round. It takes a long time, he would point out, to develop a drug. He began citing, whenever there was an opportunity, a study in the *Proceedings of the National Academy of Sciences* (PNAS) that documented the amount of time it had taken for other drugs to get from synthesis to market. EPO, the important drug used to treat anemia, had taken thirty-nine years, and the migraine drug Sumatriptan had taken thirty-five. The average length of time required to get a drug from the discovery phase to the pharmacy was seventeen years. If you thought about it in those terms, things were moving along at breakneck speed.

Some of the people who had criticized Weiner when he was riding high consoled him now that he wasn't. J. H. Noseworthy, one of the authors of the Canadian cyclophosphamide study that had dogged Weiner, sent a note of condolence, mentioning how many failed MS trials he personally had been involved in. Weiner wrote back a thank-you, saying they were on a journey together.

There was an almost valedictory tone about some of the tributes, as though Howard Weiner were about to retire from the fray. At the Cleveland Clinic, he was introduced as one of the first to advocate active treatment for MS. The speaker cited a meeting in Buffalo in 1983 where Weiner had spoken up and had been criticized by other neurologists, and another meeting on Jekyll Island in 1989, where Weiner was one of the first to talk about MS as a malignant disease. The speaker complimented Weiner on taking the heat over the years. It was, Weiner noted in his journal, "nice to hear." But even more gratifying was the news that the Cleveland Clinic, a major center for MS study, was actively using cyclophosphamide.

Ironically, just at the time when oral tolerance had taken a hit, Weiner's earlier, much-maligned idea for the treatment of MS, the use of the drug cyclophosphamide, was gaining new respectability. Weiner had long believed that cyclophosphamide was the only effective treatment for certain patients. In general, they were young patients whose MS was worsening. At the Brigham he had developed a method of infusing these patients with this drug on a regular basis that sometimes arrested or at least slowed down their MS. There were even some patients who got better. He had case histories to prove it.

One of his favorite success stories was of a sweet-faced mother of three named Cathy Crittendon who came to him soon after she was diagnosed with MS in her late twenties. Every three months or so, she would have an attack, with extreme fatigue and difficulty walking; after each attack, it would be a little bit harder to walk. Weiner tried all kinds of things to control her MS, first oral steroids, then intravenous steroids, then oral myelin, and then methotrexate. But when the attacks kept coming, Cathy reluctantly agreed to try cyclophosphamide.

Even after she decided to do it, Cathy wondered, when she began to lose her hair and had to buy a wig, if it was worthwhile. But for her, cyclophosphamide has made all the difference. "It's been like night and day," she says. "I have not had an attack in three and a half years." Now she can walk the length of two football fields without a cane or her husband's arm, and has only the beginning of a foot drag when she finishes.

At the Cleveland Clinic, researchers were having some success in treating patients who had what they called "transitional MS" and who were not responding to Betaseron, with cyclophosphamide until the disease slowed down, then returning them to the less toxic Betaseron. In July, just months after the Myloral trial failure, Weiner met with representatives of Biogen, the company that manufactured Avonex, to plan an MS trial, based on the Cleveland Clinic model, of their drug in combination with cyclophosphamide.

Thinking about the roller-coaster ride with cyclophosphamide—from all the excitement when he announced his results in 1983 to the harsh criticism from the Canadian group some years later to the recent vindication, Weiner saw a parallel to oral tolerance. "One day five, ten, fifteen years from now," he told himself, "oral antigen will have found its place in the treatment of MS."

Perhaps this new respect for cyclophosphamide, the drug that had gotten him into so much trouble in the past, contributed to the decision two months later to name Howard Weiner at last to a Harvard professorship. Or perhaps the failure of the Myloral trial and the disappointing results of the Colloral trial made the committee more sympathetic. There had been criticism from some colleagues of the fact that an article on the first human trials of oral tolerance appeared in *Science* right around the time that Weiner was out raising money for an initial public offering. But the failure of the Myloral trial made Weiner seem more vulnerable, less calculating and commercial. Also, after the trials failed, there was less to envy. Whatever the reasons, Weiner was walking to his lab one day in September 1997 when he ran into a member of the committee on professorships and learned that, after twelve years of consideration, he was at last going to be named a professor at Harvard.

The professorship saga had haunted Weiner. It was not only a personal humiliation, but a roadblock to others. David Hafler, who might be eligible for a professorship, didn't have much of a chance until his mentor was named professor. Dennis Selkoe, codirector with Weiner of the CND, had been named professor eight years before, in 1989. Each time he was rejected—and there had been at least four rejections—Weiner would remind himself that it wasn't that important. "I wish they would leave me alone," he wrote in his journal. "I will not let it turn to bitterness or anger." As time went on, even a member of the committee admitted that Weiner's case became Kafkaesque. Department of Medicine chairman Eugene Braunwald, who had been his advocate, told Weiner he had never witnessed the process working so badly in his entire twenty-four years at Harvard. Charles Barlow, head of the Department of Neurology at Children's Hospital, described it, in his letter of congratulations, as "a high-forceps delivery."

As soon as he got the news, Weiner called Robert Kroc, the McDonald's scion who had endowed the chair he held, to let him know. Kroc was ninety, and had long been indignant that Harvard hadn't given Weiner a professorship to go with his Kroc chair. Kroc told Weiner he would sleep better that night because of the news. Privately, Weiner told friends that he was furious about the professorhip: Now that he finally had it, he no longer needed it. "I was ready to die an associate professor," he wrote in his journal. But at year's end, he listed "becoming a full professor" as one of the most important events of the previous twelve months.

Meanwhile, the space issues that had aggravated Weiner for years at his lab were finally coming to an end. At the same time he learned he had been named professor, Weiner and his group finally moved to the seventh floor of the Harvard Institutes of Medicine, a ten-story building with state-of-the-art research equipment and animal rooms and sweeping views of Boston from the corner windows. After years in which lab members had to stand up along the walls for Friday presentations, the Center for Neurologic Diseases now had a huge conference room with a long oval table that could accommodate everybody with comfortable swivel chairs left over. At the first meeting there, Howard Weiner made a little joke: "We're not squeezed anymore, at least not in the conference room."

Everyone laughed. In fact, Weiner's lab had felt the squeeze from the failure of the trials. Because AutoImmune was cutting back to bare bones, the company could no longer underwrite much research at the lab. Weiner was having to look elsewhere for funds.

And even if it didn't affect their research directly, everyone in the lab was aware that, while they were living in luxury, AutoImmune, Inc. was about to move to offices in the basement of the building in which they had once been a principal tenant.

For Weiner, especially, all the academic triumphs were punctuated with stinging reminders that AutoImmune, the company founded on his ideas, was struggling to survive. Sometimes the reminders were deeply personal, as when Ahmad Al-Sabbagh was laid off. Ahmad sat in Weiner's office in tears, with his former boss handing him Kleenex. Even though Weiner might have been able to hire him in the lab, Ahmad could no longer live, with four children, on the low wages of a researcher. He needed a job in industry. Such encounters were hard for Weiner, who liked to take care of his people. "There's no question you feel the pain," he would say afterward, "but you can't be paralyzed by it."

Yet sometimes, in an unguarded moment, Weiner allowed himself to feel hopeless about things. Because of his deep attachments to colleagues and family in Israel, going there was always an emotional experience for him, akin to going home to Denver. That November, a day spent looking at old artifacts from the Dead Sea in a Tel Aviv museum led him to reflect on his years of searching for a cure for MS. "I don't know how interesting that is anymore," he wrote in his journal.

But the next day the weather was splendid. After shooting 86 during a morning of golf at Caeserea, a beautiful course between Tel Aviv and Haifa, he had lunch with Ariel Miller by the seaside. Miller, who had been a postdoc in the lab in Boston and was responsible for establishing the phenomenon of bystander suppression, had invited Weiner to the conference in Israel hoping that he would arrive triumphant, with a positive result in the Myloral trial. Now he echoed Weiner's discouragement of the day before. "You had a chance to climb to the top of Mount Olympus," he told Weiner, "but it didn't work out."

Weiner told him, "I'm starting again."

During the summer and fall of 1997, the challenge for the three remaining members of top management at AutoImmune—CEO Bob Bishop and Vice Presidents Jo Ann Wallace and Malcolm Fletcher—was finding a way to hang onto the small band of surviving employees while they, and the AutoImmune board, decided on the future course of the company. "If we're to go anywhere with this thing there are a few people we've absolutely got to keep," Malcolm Fletcher noted.

"The goal is to keep enough light at the end of the tunnel to maintain hope," Bob Bishop explained. "I think the people remaining here have hope, but they're also scared."

Still, in an excellent job market, the company couldn't rely on hope alone. So they put together a package that included an improved deal on stock options and a severance payment so high that it encouraged people to stay on until they were fired. Then, on the Friday before the Fourth of July weekend, Malcolm Fletcher gave a talk about the data from the trials that was designed to lift people's spirits.

Since it was casual Friday, Malcolm stood before the twenty or so employees wearing a sweatshirt and jeans, which made him look even more boyish than usual. "I sort of enjoy this data," Malcolm told the group, as he began to put overheads up on the screen. Everyone laughed.

"I'd like to announce," Bishop chimed in, "that Malcolm is a very perverse person."

Reaching up under his sweatshirt to extract a pen from his shirt pocket, Malcolm began his exercise in optimism. "There are ways of looking at this material," he told them, "that will give us a label." He then proceeded, using transparencies, to point out the ways in which the Colloral studies, taken cumulatively, were promising. He also noted that there was still one ongoing study: a comparison of Plaquenil to 20 micrograms of Colloral. Asked what he would do if that were positive, he told the group: "I will be at the FDA like a rat up a drain."

But privately, Fletcher and others didn't hold out much hope for the Plaquenil trial, because the Colloral in that trial was being given in the 20-microgram dose, which had been demonstrated to be too low. It was one of the many sources of aggravation that were causing friction among the officers of the company. For a time that summer, Jo Ann Wallace and Bob Bishop weren't even speaking to each other.

In August, at one of the lowest points in the company's journey, Wallace, while watching the U.S. Women's golf tournament at the Braeburn Country Club near Boston, fell into a hole and fractured her leg in two places. There was another "defection" at the company, bringing the payroll down to seventeen. And AutoImmune finally got the results of the 009 trial: Colloral wasn't as effective as Plaquenil. Nothing was turning out right.

What most of the survivors wished for was that AutoImmune would decide to proceed with a phase 3 trial of Colloral, ending in success and a drug on the market. But it wasn't at all clear in the summer of 1997 that this would be a wise use of the company's remaining $35 million. Bishop estimated the chances of success in a phase 3 Colloral trial were 50–50. That just wasn't good enough.

There was one last hope: the possibility that a metanalysis of the results from all five phase 2 trials might provide stronger evidence than any single trial for going forward. So AutoImmune hired Boston BioStatistics, the company in Framingham that had served them well before, to undertake what is called an "integrated analysis." If the data were positive, AutoImmune might go to the FDA and propose a phase 3 trial or even—wildest of dreams—ask the FDA to approve the drug right away.

In the interim, the AutoImmune board considered a range of possible uses for AutoImmune's remaining $35 million. There were investment bankers who wanted the company to abandon oral tolerance altogether—to go shopping with AutoImmune's $35 million for another technology. "We could take the money we've got," Bob Bishop explained, "and start to ride a new horse."

But Bishop was proud of the fact that he'd been smart with AutoImmune's money, making painful but necessary cuts. He wasn't about to use it now to rescue another fledgling outfit that hadn't planned as well. Another possibility, perhaps preferable, would be to simply return the money that was left to the shareholders. Or the company could go back into research and get very, very small, giving up its manufacturing activities.

Fortunately, the integrated analysis of the Colloral trials provided another alternative. It turned out the pooled data from all the phase 2 studies, which involved more than 1,200 patients, showed Colloral to be significantly more effective than placebo, especially in the 60-microgram dose. The chances for success in a phase 3 trial were considerably better than 50–50.

By September, the company had a plan. A letter, signed by Bob Bishop and Barry Weinberg, chairman of the board, went out to shareholders. The focus would be on gaining approval for Colloral as soon as possible. Sometime around Thanksgiving, the company would go to the FDA to report on their phase 2 results and to seek approval for a phase 3 trial. "It is our strong belief," the letter concluded, "that oral tolerance has a bright future and represents a viable therapeutic approach to autoimmune disease. We look forward to completing the development of Colloral, as well as our other products, and are committed to seeing that AutoImmune's value is recognized by the investment community."

Jo Ann Wallace had recovered enough from her leg fracture, and the surgery that followed it to return to the office at the end of October, in time to oversee the planning for the presentation on December 15 to the FDA. It was, she knew, their only chance. "It needed to be perfect." Wallace insisted that they put together a substantial presentation. "You load them up with data, so they don't think you're hiding things or cherry-picking what you've got." She suggested that Joe Boccagno, a handsome recruit from Johnson and Johnson who had come on board as senior clinical trials manager, do the presenting. Malcolm Fletcher, she noted, is "a global kind of guy" who would probably try to present too much. Joe, on the other hand, was "very buttoned down" in the J and J tradition. Malcolm Fletcher would attend, of course, as "head of state," along with J. D. Bernardi, AutoImmune's director of regulatory affairs, and Phil Lavin, the statistician from BioStatistics who had done the integrated analysis. They also recruited George Ehrlich, a distinguished rheumatologist from the University of Pennsylvania, as well as another statistician from Duke, Gary Koch. Bob Bishop wanted to go too, but Wallace convinced him to stay away. She was worried that his enthusiasm might undermine the company's credibility with the FDA scientists. She planned to stay away too, for similar reasons. "I'm a marketing person," she noted, "that can tick some people off."

"We worked on the presentation day in and day out for three weeks," Wallace remembers. "We rehearsed every day, once a day, until we finally got it down. And we went down the day before, checked into the Holiday Inn in Bethesda, and rehearsed, over and over. When we got done rehearsing, I said, 'guys, we're ready. Kill 'em.' And they all got in a cab and went over to the FDA."

During the FDA presentation, which was scheduled for two, Jo Ann Wallace waited in the USAir Club at National Airport. Not long after five, she saw two of the team, Phil Lavin and J. D. Bernardi, running in her direction. It was, they reported, a "great meeting." The FDA panel gave a green light to a phase 3 trial. And because Colloral is a uniquely safe drug, they agreed that a single phase 3 trial, if successful, would be all that they required for licensure of the drug. Had Bob Bishop been there, he would have called it a home run.

February 7, 1998, Conference Room, Harvard Institutes of Medicine

The researchers in the Weiner lab are gathering around the long oval table in their new quarters, waiting to hear from Joe Boccagno, the recent AutoImmune recruit who is going to talk about the phase 2 and upcoming phase 3 Colloral trials. As usual, there are copious breakfast foods laid out in the anteroom, and people are loading up with bagels, muffins, juice, and coffee before settling in to listen. Joe Boccagno brings news that is important to them, since it grows out of the clues they are extracting, slowly and painstakingly, from the living cells that are their daily preoccupation. But at the same time, it is news from another world: the world of clinical trials, and patients, and Wall Street. Quite a change from the usual Friday sessions.

Up front, Joe Boccagno has his overheads ready to go. Unlike Malcolm Fletcher, with his tendency to range all over the place in charming fashion, Joe Boccagno is an efficient but unexciting presenter. His looks, however, are dazzling: He is dark and handsome enough to be a magazine model. His black hair is brush-cut with just a little gray at the temples, and his very dark eyebrows are bushy punctuation marks on his tanned face. The son of an army officer who grew up on bases around the world, Boccagno's grooming and posture are military. Among Howard Weiner and his scruffy crew, such style is rarely witnessed.

Some of the usual faces are arrayed around the large oval table. Howard Weiner, always sitting right down front, near the speaker, and, farther back, Ruthie Maron and Anthony Slavin. But there are new faces as well. Gabriela Garcia, the gentle Brazilian who worked closely with Ruthie and Anthony, has moved on from studying rheumatoid arthritis in mice to a job in private industry. A new Brazilian, Ana Faria, has come on in her place. Other new people are there not so much because of downsizing, as at AutoImmune, but because of the natural ebb and flow of the post-docs in the laboratory.

Boccagno begins with a brief reference to the Myloral trial—a gamble that didn't work out. Then he goes on to the good news. In December, he and others went to the FDA to propose a phase 3 trial of Colloral and "the agency" gave their blessing. If the trials demonstrate that rheumatoid arthritis patients have a significant reduction in three of the "core four" categories—swollen joints, tender joints, the physician's global assessment, and the patient's global assessment—the FDA is pretty much guaranteeing that they'll license Colloral.

AutoImmune is putting together a large trial. The plan is to recruit 800 patients and to make the entry criteria stiffer than in the past. Patients must have at least twelve tender and ten swollen joints to qualify, and if they are on disease-modifying drugs, like Methotrexate, they will have an eight-week "washout" period before they begin the trial. Sometimes, drug trials allow more than half of the trial patients to be on drug, as opposed to placebo, but this time the divide will be half on drug and half on placebo.

The reason for all these decisions—the large number of patients, the high level of disease required, the high percentage of patients who will be on placebo—is that they improve the chances of success with a drug that has a mild effect at best. If Colloral were a blockbuster drug, you wouldn't need to have such large numbers or such serious disease to show efficacy. And you wouldn't need to put half the patients in your study on placebo to tease out the difference between the active and placebo effects. "We're doing a large enough study to show us if Colloral is effective," Joe Boccagno concludes. "If we fail, then Colloral is dead."

In the discussion that follows, Howard Weiner quotes a proverb he has learned from Kostya Balashov, the Russian postdoc in the lab. "A positive result," the saying goes, "is a gift from the devil." It is an unusually gloomy comment for Howard Weiner to make, and it is a hint of the private worries he has expressed in his journal as AutoImmune has moved toward the decision to do a phase 3 trial. Even as late as November of the previous year, when the company was about to commit to it, he wondered in his journal if it was the best option. He worried about "hitting the wall at 100 miles an hour." But now it is too late to turn back.

As the session ends, David Hafler suggests that the lab members take a vote on whether the trial will succeed. Little pieces of paper are torn up and passed around, everyone writes down their best guess and drops it into a basket at the door. Hafler counts the votes and announces the result: there are eighteen yesses and nine nos.

CHAPTER 13

The Rheumatologists

MARTHA BARNETT, ONE OF THE RHEUMATOLOGISTS involved in early trials of oral tolerance as a treatment for rheumatoid arthritis, talks fast and moves quickly through her busy life, which includes young children, seeing patients, and juggling schedules with her physician husband. But she remembers a dream she once had in which she understood, for the first time, how different life would be if she were diagnosed with the rheumatoid arthritis she often treats. In the dream, every movement was difficult. She realized that she would have to take pills every day for the rest of her life, and return over and over to the doctor's office for exams and blood tests. She woke up in a cold sweat.

Barnett, who has blonde-streaked hair and wholesome good looks without makeup, was a rheumatology fellow when she had the dream. She had been seeing rheumatoid arthritis patients for a while by then, dispensing medications and listening to their complaints. But after she lived the experience in the dream, she understood rheumatoid arthritis (RA) in a new way. "Somehow it all came into focus that it affected *everything!*"

RA, like MS, is a stealth disease. It comes and it goes, often in synchrony with the weather. Those who have it are sometimes misunderstood by those around them for that reason. "The hardest part," as one RA sufferer puts it, "is that it's not constant. When I can't do things, it's, 'well, you could do that the other day, how come you can't do it today?'" Because it affects the joints, particularly in the hands, arms, legs, and feet, RA interferes with all the activities required to get through the day: getting out of bed, buttoning buttons, putting on socks, turning a doorknob, getting up and down stairs, opening a jar, turning the steering wheel of a car. Lifting a child or putting a plant in the ground is sometimes impossible, particularly during flare-ups. Working on a regular basis becomes difficult.

In the past, severe RA consigned its victims to wheelchairs, with their joints frozen and twisted beyond use. Now, because of joint surgery and a host of medications, there are ways to keep most rheumatoid arthritis sufferers mobile. But the pain and fatigue of the disease persist. "I've seen patients whose lives just get chipped away," Martha Barnett says. "They keep trying. But they give up a little bit of this, and they give up a little bit of that and their life just isn't what it was."

Thomas Sydenham, the seventeenth-century physician sometimes called the "English Hippocrates" because of his acute powers of observation, described the course of untreated rheumatoid arthritis well. "It will last for months and years," he wrote. "Nay, it will torment a patient throughout his miserable lifetime. Its violence indeed may vary: so that, after the fashion of gout, it may come on at odd times, and in periodic fits." Even after the pain ceases, Sydenham declared, "the Patient shall be a cripple to the day of his death."

Though he delineated the symptoms, Sydenham and other physicians of his time lumped RA in with other joint inflammations under the broad rubric of gout, the disease du jour. Sydenham himself was a gout sufferer, and noted that his was an ailment that "kills more rich men than poor, more wise than simple. Great kings, emperors, generals, admirals and philosophers have all died of gout. Hereby Nature shows her impartiality, since those she favors in one way she afflicts in another."

Because it was a disease of the privileged in an age of class distinctions, gout was a preferred diagnosis. "This seems to be the favourite disease of the present age in England," the physician William Heberden observed in 1802, "wished for by those who have it not, and boasted of by those who fancy they have it, though very sincerely lamented by those who in reality suffer from its tyranny." Heberden was one of the first physicians to suggest that certain symptoms differed so much from typical gout that they deserved to be called by a different name. He described "an articular affection, not commencing in the great toe, but preferring other joints." When the "attack" began in other joints, he noted, "the continuance of such a pain, the return of it, and its consequences, will differ so much from the ordinary gout that it is either to be called rheumatism, or should be distinguished by some peculiar name from both of these distempers."

The "peculiar name" that stuck was assigned fifty years later, by Sir Alfred Baring Garrod, another eminent British physician. It was Garrod who first identified the cause of gout, which is brought on by an excess of uric acid in the blood and can be caused by an over-rich diet. Although "unwilling to add to the number of names," Garrod suggested that "rheumatoid arthritis" might be used to describe "an inflammatory affection of the joints." Garrod went on to describe a number of the features that modern practitioners use to diagnose the disease. The inability of patients to "stir in bed on first waking," commonly referred to as "morning stiffness," the swelling of joints owing to the "presence of fluid in the cavities," and the "twisting outwards of the fingers" in advanced disease. In his 1859 treatise, Garrod included a drawing of a hand "much distorted from rheumatoid arthritis." The wrist is swollen and fingers of the hand are rigid and bent outward, in the direction of the little finger.

Garrod's name, rheumatoid arthritis, helped to distinguish the malady from gout, but it called up new associations that continue to cloud the picture. Rheumatoid arthritis has very little to do with rheumatic fever, despite the overlap in the names. Rheumatoid arthritis is also an entirely different disease than another much more common kind of arthritis, osteoarthritis, which afflicts most of us to some degree if we live long enough.

One of the most frequent challenges of a rheumatology practice, according to Martha Barnett, is making the distinction between rheumatoid arthritis, which is a systemic illness, and osteoarthritis, which is mechanical in origin. In osteoarthritis (OA) there is pain in the joints but little swelling. RA, on the other hand, is caused by an inflammation of the synovial tissue that lines the joints, resulting in a "big mass of inflammatory tissue that's wearing away cartilage and bone. Sometimes you can't see it, and you have to just get down and feel." There are other distinctions as well: RA, because it is systemic, tends to occur on both sides of the body at the same time, whereas OA does not. RA occurs in younger people—indeed, there is a juvenile form of the illness. RA activity can sometimes be measured by a marker in the blood. But even with all these differences, it isn't always easy to tell which disease is causing joint pain. Yet it is absolutely crucial to distinguish between RA and OA. "You've got to pick," Martha Barnett explains. "It doesn't do any good to sit on the fence because you have to do different things about them."

Fifty years ago, rheumatologist Lea Sewell notes, there wasn't much reason for the field of rheumatology to exist because there was little to do about the diseases, including RA, that a rheumatologist might treat. "Why have a specialty for something you can't do anything about? Arthritis a specialty? That's like itching your ear should be a specialty."

That began to change with the stunning discovery of a Mayo Clinic doctor named Phillip Hench. Hench was an optimist if ever there was one, and he bemoaned the "gloomy tradition" among his colleagues of looking upon rheumatoid arthritis as an untreatable disease. It was true, he conceded, that "rheumatoid arthritis has come to imply more or less relentless progression to crippledom." But rather than focus on the exacerbations of the disease, he chose to study periods when the disease went into remission. He noticed that women tended to have remissions during pregnancy and that patients with jaundice also went into remission. This led him to doubt the then-current assumption that RA was caused by a microbe. "It has become easier for me to consider that rheumatoid arthritis may represent, not a microbic disease, but a basic biochemical disturbance." This was a paradigm shift that would make it possible, within a few years, for Hench and his colleague, the chemist Edward Kendall, to make a spectacular discovery.

If RA was, as Hench suggested, a "biochemical disturbance," then biochemistry might provide a cure, what he called "nature's dramatic antidote," the antirheumatic "substance X." He first tried the obvious: He gave hormones like those activated in pregnancy to some of his RA patients, to see if they helped. He even induced jaundice in a patient with severe RA, hoping it would be beneficial, but neither of these strategies seemed to have any impact whatever on the disease.

Then in 1949, in a preliminary report at the staff meetings of the Mayo Clinic, Hench announced that he had found his substance X. It was an extract from the adrenal cortex, the outer portion of the adrenal gland, and it had first been used successfully in 1929 as a substitute in patients whose own adrenal gland had stopped functioning. "Compound E," as the substance was called in the beginning, had been a prospect for a long time before it was tried. But it was hard to come by, mainly because every effort in the mid-forties was concentrated on the war in Europe and the Pacific. Once Hench was able to obtain it, and try it out on rheumatoid arthritis patients, he knew immediately that he was onto something that would transform the field. "Compound E," renamed cortisone, was a marvel. "It is not too much to say," as one chemist commented at the time, "that a revolution in medicine was in preparation."

The first case Hench treated with his compound E was a twenty-nine-year-old woman who had had severe rheumatoid arthritis for over four years. In September 1948, her disease had worsened to the point at which "she could hardly get out of bed." Hench gave her 100 milligrams of compound E by injection, but observed no change that day. "Walking was so painful," he reported, "that she ventured only once from the room." Nothing much happened on the second day either. But on the third day, "she rolled over in bed with ease, and noted much less muscular soreness." And on the fourth day, her painful morning stiffness was entirely gone and she walked with only a slight limp. On the seventh day after the shots began, the patient "shopped downtown for three hours, feeling tired thereafter, but not sore or stiff."

It didn't take long for excitement about this new discovery to build. In 1950, Hench and Kendall, along with a Swiss chemist of Polish origin named Tadeus Reichstein who contributed to the isolation of the adrenal hormones, won the Nobel Prize in Medicine.

Phillip Hench sparked a transformation of the then-fledgling field of rheumatology. Now there was something to do for RA patients. Hench's work pointed to the fact, which would be elaborated by others, that RA was a systemic illness, caused by what he called a "biochemical imbalance" in the body. The field of rheumatology grew to include 110 different "connective tissue" diseases, including lupus and psoriasis among others, many of which are now understood to be autoimmune diseases, caused by a reaction of cells within the body against the body's own tissue.

In his Nobel acceptance speech, Hench referred to cortisone and ACTH, another form of the substance, as "unframed pictures," needing clearer definition. Here, he was more right than he knew. Although the corticosteroids, as this whole group of medications are called, continue to be crucial remedies for inflammation in a host of conditions, including MS as well as RA, their side effects make them a mixed blessing. No one, nowadays, would think of giving a patient a 100-milligram dose of cortisone or of prednisone, which is most commonly used in RA.

"Almost every patient responds to prednisone," rheumatologist Lea Sewell notes, and adds that "if you took 40 or 60 milligrams a day, your arthritis would be pretty quiet." Unfortunately, however, the corticosteroids have life-threatening side effects, particularly in large doses: They suppress the body's immune system, leaving the patient vulnerable to infection, and they can cause bruising and thinning of bones over time, as well as cataracts, weight gain, a round "moon face," diabetes, and high blood pressure. When corticosteroids were first discovered, rheumatologist George Ehrlich notes, everyone thought, "we have a cure." But it didn't turn out that way. "They don't work all the time, they produce side effects, and they're not cures, all they do is suppress symptoms."

There have been other breakthroughs in the treatment of rheumatoid arthritis since Hench's discovery of cortisone. Some point to methotrexate, a cancer drug that turned out to be very effective for many RA patients, as a milestone. Most would agree that the new drugs introduced in the last few years have accelerated the pace of progress dramatically. But you only need to spend a few days with Lea Sewell, a rheumatologist and researcher at Harvard's Beth Israel Deaconess Medical Center in Boston, to understand not only how far the treatment of RA has come but also how much further it has to go.

It's a sunny Tuesday in August, and Lea Sewell swoops into the Souper Salad restaurant at the Beth Israel, twenty minutes behind schedule, to grab gazpacho and lemonade to go. "Too many prescriptions to write," she explains, as she leads the way, white coat open and flying, to her research office in the basement of one of the old buildings on the campus. She has five minutes to get to a meeting with a representative from Amgen about a new arthritis study, so she walks even faster than usual. Amgen is proposing a phase 3 rheumatoid arthritis trial of Interleukin 1-antagonist, an anti-inflammatory cytokine they've been working on for some time. Sewell likes what she knows of the study, and is eager to get a green light from the company to recruit patients. "Recruiting is the name of the game," she notes.

As Sewell speeds along, high color intensifies in her cheeks and her thick, shiny black hair bounces with her stride. She wears skirts to work, underneath her three-quarter-length white coat, and low heels, scuffed from a lot of walking. Two plastic ID cards, one on top of the other, are clipped to her collar and the words Beth Israel Deaconess are embroidered in blue on a patch underneath.

Lea Sewell had always thought when she was growing up that she would wind up in biology. But two decades ago, as an undergraduate at Dartmouth, working in the lab of a senior researcher, she noticed how much time was taken up with writing grants and how long it took to get results. "I think I wanted more successes every week, not two or three a year for a lifetime."

More and more after that Sewell thought of becoming a doctor.

She remembers walking around the Dartmouth campus one snowy afternoon trying to decide what do. "I thought, 'Can you be a woman and a wife and mother and a doc?' That was my only question. Because I was unwilling to give that up. And I said, 'Yes, I can do that.'"

Yet laboratory science still intrigued her. During medical school at Duke in the early eighties, she spent enough time in the lab to become fascinated with the remarkable new developments in immunology. One of the attractions of rheumatology, for her, was that it meant grappling with the newly discovered complexities of the immune system. As Sewell's friend and colleague Martha Barnett notes, medical specialties are arrayed along a continuum. "People are either more interventional or more cognitive." Surgery is at the interventional extreme of the continuum. If you choose rheumatology, where the immune system is your territory and where the answers tend to be complex and incomplete, you have chosen the cognitive end.

Sewell's current position, part researcher, part staff physician, is an ideal melding of her scientific interests and her wish to take care of patients. What she does is sometimes called "translational" research, because it provides a bridge between basic science and clinical application. But even though she likes the niche she's in, Lea Sewell is often unhappy on the job. In fact, it sometimes seems as if there are two Dr. Sewells. When she is seeing patients, Lea Sewell couldn't be more sympathetic and patient. When she squeezes a joint and the patient winces, Sewell says, "sorry." But when she's dealing with colleagues and with the bureaucracy, Sewell is often impatient or enraged. Like most doctors these days, she must spend a lot of time thinking about how to justify the cost of what she does. And she thinks the system undervalues doctors in general, and clinical research in particular.

Right now, Sewell laments, doctors are "viewed as a cost center not a profit center" in the organization, even though they are the "cash cows." Physicians, she complains, are under constant pressure to justify all their costs. But no one gives praise or credit for what they do to ensure the quality of care. "There's no one saying, 'Oh, you did a good job, you got an X ray for that patient.' There's only the negative."

The one thing that gives Sewell some autonomy and fiscal clout within the increasingly burdensome bureaucracy is clinical research. The amount of money paid by drug companies to clinical research sites is a closely held secret. Jo Ann Wallace, of AutoImmune, refuses to divulge any figures. One reason for this is that the payment, which is made on a per-patient basis, varies from one place to the next. "The range," Wallace acknowledges, "is huge." Companies don't want one site to know what the other's getting. According to a 1999 report on clinical trials in the *New*

York Times, the payment to sites can range from $1,000 to $4,410 per patient. The drug company, according to Lea Sewell, "has a maximum in mind, and will not use a clinical site if they go over that number." But there are factors that raise the value of a site in the company's eyes. A "thought leader" like Harvard can demand more. It also helps if the investigator is considered to be a "good enroller," particularly if the disease in question is rare, or if previous studies have depleted potential candidates.

Lea Sewell, situated at a Harvard teaching hospital that is a magnet for patients with unusual diseases, manages to attract a lot of drug company research at a decent per-patient fee. Even after she pays out a large percentage of the money to cover office space, laboratory, and administrative costs, Sewell manages to support a part-time research nurse and to augment her own salary from her clinical research. That is why it is important to have a successful meeting, on this day, about the new Amgen trial.

The Amgen representative is sitting in a corner of the crowded research office, chatting with Michele Finell, when Sewell arrives. She has come from Amgen's headquarters in California to look at a number of sites in the New England area, and discusses the entry requirements for the fifty-two-week trial with Sewell and Finell. They have some questions and suggestions for her, which she promises to take back to headquarters. Then she asks for the customary tour of the facilities where the trial will take place.

The Amgen representative is young, with a B.A. in psychology and limited experience at her job, but Sewell and Finnell, who might be annoyed with her under other circumstances, treat this particular neophyte well. They devote perhaps an hour to giving her a tour. She takes careful notes, and they show her two labs on different floors in the hospital, walk her past radiology, then take her up to the research unit, a floor of the hospital where patients with severe RA are involved in an inpatient test of a new drug. They introduce her to the pharmacist, and show her the pharmacy from which trial drugs are dispensed. Then, because she is required to see every space used in the trial, they walk with her over to the building where Sewell has her rheumatology practice and show her those labs and offices as well. Before the rep leaves, Sewell and Finell have what they want: a tentative date in September for the pre-trial investigator meeting in Arizona, and enough information on the requirements and exclusion criteria to begin looking for candidates for Amgen's phase 3 trial of a new arthritis drug called Anakinra.

The next day, Sewell sees a patient she believes is an excellent candidate for the Amgen trial. She is a young-looking forty-seven-year-old black woman, Rita M. who was diagnosed with RA four years ago, and whose disease does not yet show. For a long time after she got the diagnosis, Rita kept expecting to get better. "I was in big denial," she acknowledges. "I kept telling the doctor, 'I'm not bouncing back!' She looked at me and said, 'You have rheumatoid arthritis!'"

In fact, Rita has what Lea Sewell calls "a lot of arthritis," which is reflected in the sedimentation rate in her blood, and she lives with constant pain. An operating room nurse for twenty-five years, she had to quit her job because of the RA and go on disability. These days, she volunteers in an Italian home for foster children and pays attention to her illness, swimming several times a week if the water is warm enough.

Lea Sewell tells Rita about the Amgen trial, explaining that she would be allowed to stay on her methotrexate, the strong drug she's taking, but would have to give up the milder Plaquenil. The Amgen drug would require her to give herself "a little needle shot." Sewell does not mention that fact that there is a chance Rita will be on placebo during the trial. But Rita, who has been in trials before, probably understands that. If and when she decides to participate, there will be a discussion of placebo and other details of the trial previous to her giving "informed consent."

Sewell sits on the edge of her chair, arms folded over her chest, bare legs crossed at the ankle, as she explains the protocol. Occasionally, she unfolds her left arm, but keeps it close to her body, moving her open palm up and down for emphasis.

Rita's beautiful dark eyes are intent as she considers the possibility of entering the study. She has been in others, including the trial of a drug that seemed to be helping her, before it was withdrawn for safety reasons. She is open to new possibilities, though wary of introducing foreign chemicals into her body. This is a concern she shares with many other patients when they are confronted with the possibility of entering a trial. Sewell tells Rita that the Anakinra is "something that bodies make." She also mentions that the Amgen product may help prevent damage to bone.

This is something that is bound to appeal to Rita, whose attention to her body is obvious. There are diamond bands on her pinky and fourth fingers, and she wears gold bracelets and a gold bracelet-style watch. Her fingernails, polished in red, match her toes. "I sleep with my hands spread out under the pillow," she confides, "because I don't want them to get deformed." Rita agrees to take home the literature on the Amgen trial and give it serious thought. "One door closed," she says, referring to the previous trial, "but another has opened."

The next morning, Sewell tells Michele Finell, when they meet to go over the various ongoing trials in the research office, that she suspects Rita will agree to go into the Amgen trial. The two of them discuss the logistics of the trial, which require "blinded joint counts." That means that an additional trained person will be needed to examine the joints of patients in the trial to assess their swelling and tenderness, keeping a record that Sewell and Finell will not see. A young doctor, who has been assigned to Sewell and has been sitting in on the discussion, agrees to take on the job of assessing joints. But then it turns out that she has just begun her fellowship in rheumatology and has never done joint counts. "Go home and feel everything on your husband," Lea Sewell tells her, with a flicker of annoyance. Then she has a better idea. "Go spend time with Betsy. Betsy will teach you."

Betsy M. is an RA patient who is currently spending the week in a room on the research floor of the hospital, taking part in a continuation trial of a new drug called Ontak, aimed at providing relief for severe RA. Ontak was pioneered by a Boston biotech company called Seragen, notorious in local circles because it was propped up through hard economic times by an enormous investment from Boston University under the leadership of then-president John Silber. The company lost money for years, and so did BU. Critics said Silber had exposed BU contributors' money to too much risk. According to Lea Sewell, however, the story has a happy ending for patients, if not for all investors. The drug Ontak, which is used in certain cancers, also turns out to be effective in some severe cases of RA, although it has yet to be approved for this use by the FDA. Betsy M., who has been helped by it before, is hoping it will provide relief for her during her current flare.

Betsy M. is now forty-seven, and has had RA since she was seventeen. She remembers the first time she felt numb "from the neck down." It was in 1977, and she was with her husband, who was in the military until recently, in Guam. "I'm a joker," she says, "so when I told my husband I couldn't move, he thought I was kidding. But I wasn't."

Betsy M., whose thin face accentuates the prominence of her eyes, is leaning back on her patchwork pillowcase, brought from home "because everything here is so drab." A whole wardrobe of T-shirts, including one that says "Out of body, back in five minutes," are lined up on hangers beside her bed. Today, she's lying on top of the bedcovers in gray sweatpants. Her feet are bare.

"I've had a lot of work on my feet," she says, looking down at them. "Those are fake toes." Betsy M. has had both big toe joints replaced and her feet have shrunk in the process from a size 8 to a size 6. She has had two hip replacements, one knee replacement, one elbow replacement, and replacement of all her knuckles on both hands.

Even when she's not flaring, Betsy M.'s hands are bent sideways at each joint and stiffened by RA, with bulges of cartilage at the wrists, and scars across the knuckles where the surgery has been done. Right now her fingers are red and swollen. The disease, plus the surgeries, have left her with no sensation in her left hand. And since she is left-handed, she cuts herself and doesn't know it until she sees the blood spurting out. "My family gets mad," she says, "when they see me using sharp knives or getting near the oven." To add to her worries, her daughter, who "never knew what it was like to have a healthy mother," has struggled herself with a rare autoimmune illness called Marfan's syndrome, which can adversely affect the eyes and the heart. Also, not long ago, her husband, who does powercleaning for businesses, had a driving accident that put him out of work. The family has no insurance.

On this particular morning, though, Lea Sewell is bringing some very good news. Betsy M. has been trying for some time to get a waiver from the hospital so that she can get free care. This morning, Lea Sewell arrives with a piece of paper that grants it. Betsy M.'s eyes well up. "Oh my God, I'm gonna cry!" she says, and gives Sewell a hug. "The woman I know down there in free care told me you might be able to do it, because you're a doctor."

Lea Sewell and Betsy M. chat for a while about Betsy M.'s family. Her daughter is improving and her husband has found some work. Before long, though, Sewell's beeper goes off; a patient is waiting at the office in the other building. She leaves the ward, grabs some takeout noodles from Souper Salad, and walks briskly out the front door of the hospital clinic in the direction of her office. "There are so many uninsured," she says, thinking of Betsy M., "at the upper-blue-collar level. Gas station attendants who make too much money to qualify for Medicaid. And there are a lot of young uninsured people too, kids who leave the family plan and figure they're healthy and don't need insurance. But then they go skiing!"

The examining room where Sewell sees patients is decorated, sparsely, with the gifts of drug companies. There is a Dufy poster, courtesy of Merck, on the wall, and a very convincing model of a hip joint, provided by Naprocin. The blue upholstered chair, with its lumbar curve, still has a tag hanging from it. It is an unloved space.

But down the hall, in the small office to which she retreats when she's not seeing patients, Sewell has added some personal touches. On the worn oak desk there is a picture of her with her new husband and various family members. An Amish quilt, with a surprising juxtaposition of black with pink, red, and turquoise covers one wall. The round table with her computer on it looks out over the lush green of Olmsted's Emerald Necklace, a long swath of grass and trees that skirts the Longwood medical area.

Often, though, the serene view out the window is at odds with the atmosphere within. Between patients, Sewell calls someone about a mistaken charge on one of her research grants, which was supposed to be corrected a long time ago. Someone on the other end of the line is explaining that they haven't had a chance to fix it because things have been so busy. Sewell's whole face takes on the red that is usually confined to her cheeks. "Do you think you could have called me, instead of waiting a month?" she asks, her voice going lower in her fury. She hangs up. "Our research accounts are trash," she says, with a wave of her hand. Then her anger seems to dissipate, as she moves to respond to the beep that tells her that her next patient has arrived.

The patient is a professional woman in her fifties who has a lot of the symptoms of RA, although it isn't showing up in her blood. She describes being stiff in the morning, and getting very tired and slightly feverish in the afternoon, and when Sewell examines her she has quite a bit of swelling and pain in her joints.

At first, the patient seems calm and matter-of-fact about her new illness. But she has brought along her partner, a woman of about her age, and the partner is clearly worried. The partner wants to know, first of all, could it be cancer?

This is a cue for Lea Sewell to go into reassuring mode, which she does extremely well. Earlier in the afternoon, a forty-nine-year-old woman had come to see her with a list of symptoms, written on a small piece of crumpled paper that she pulled from her handbag. The patient had pain around the eye, a shooting pain in her head, traveling joint pain, a tingly feeling, an itchy vagina and anus, and hands that burned when she typed. Furthermore, she wasn't sleeping because she was anxious about the pain. Lea Sewell listened patiently, taking note of the special diet and vitamin supplements the patient was on, and did a careful examination. Then she stood before her, hands on hips, and pronounced, "I see no sign of rheumatic illness."

"You're not worried at all?" the woman asked. Her face was damp with perspiration. "I'm not worried at all," Sewell answered. The woman left, looking tremendously relieved.

Now, in response to the cancer question, Sewell is again clear and reassuring. "It's not cancer," she says. And furthermore, "arthritis is a nag and a pain in the butt, but it's not fatal."

The partner goes on to say that the patient is a lot less calm than she is acting in front of her doctor. In fact, she is "very discouraged, very teary and depressed." She's also dizzy at times, and her gait is uneven.

Now the patient begins to open up. She talks about how upsetting it is to realize that, when she can't do things, "people think I'm faking." And she talks about her frustration, after a lifetime in which she has always had a lot of energy. "I can't stand not being able to do what I want to do," she says.

Sewell explains that the early stages are often the hardest, because "you have zero medication and zero control." There is still a possibility that the RA will go away, especially since there's no RA marker in the blood. But meanwhile, Sewell tells her, she will have to accept some changes in her way of living. She needs to take time for lunch, rather than eating on the run, and she has to expect that she will get tired.

"That's just it," the patient says, with tears in her eyes. "I want my life back!"

After the couple leaves, Lea Sewell ruminates about the patient's illness. She has a lot of inflammation, but she's seronegative (has no marker in her blood indicating RA). It may be, Sewell explains, that this patient has a different disease than those who are seropositive. Her arthritis could be postviral, as with the arthritis in Lyme's disease. "Seronegative RA may be one of a group of diseases that we haven't separated out yet," just as Lyme's disease once was. In this way, the process of distinguishing and defining continues. Just as gout, thought to be one illness, turned out to be many, RA is probably more than one disease. As in the past, making distinctions is an important part of finding better treatments. "That's why," Lea Sewell notes, swiveling around to her computer to write up the case, "we do research on this disease."

CHAPTER 14

Two Meetings

March 1998, Charleston, South Carolina

The FDA-required meeting between clinical investigators and the drug company at the beginning of a trial has been known to turn into a shouting match. So it shouldn't have come as a surprise to the executives of AutoImmune that the first meeting with investigators at the start of the phase 3 Colloral trial was, according to one of the doctors in attendance, "pretty much of a free-for-all."

Things began pleasantly enough, on the evening before the day-long orientation, as rheumatologists and coordinators (sometimes nurses, sometimes office administrators) from all over the United States flew into town. There were teams from Nevada, from California and Oregon, from the South and the Northeast, representing nearly half of the thirty-six sites in the trial. (A second orientation meeting was to be held later, in Las Vegas.) For most of those in attendance, this was a first experience with Colloral. But both Lea Sewell and Martha Barnett, who was now practicing in Virginia, were there, preparing for the decisive test of the drug they had helped to develop.

The Carolina air was balmy, and women selling sweetgrass baskets were out on the steps of the historic buildings along Meeting

Street. There were white camellias and fragrant jasmine in bloom everywhere. The accommodations, in the Charleston Place Hotel on South Market Street, with Gucci next door and Saks across the street, were deluxe. The evening before the meeting, AutoImmune hosted a horse-and-carriage ride to a delicious meal at a good restaurant.

The next morning, Malcolm Fletcher started off the meeting with an attempt to leaven the atmosphere. "I've got this dangerous equipment on that's supposed to make you able to hear me," he began, as he adjusted the wireless mike on his shirt. Fletcher was in shirtsleeves, and wore a yellow-and-blue tie that, as Jo Ann Wallace put it, could "cause someone to have a seizure." His humor didn't seem to amuse the investigators, who were seated at long tables under large, glistening chandeliers, with several pounds of paper in three-ring notebooks on the white tablecloth in front of them.

A clinical trial requires close cooperation between the drug company and the clinicians, who watch over the subjects. But the doctor/company alliance, although necessary, is a tricky one right from the start. There is, first off, the issue of money: How much will the clinicians at the trial site be paid? Everyone knows this per-patient sum varies widely, and that leaves open in people's minds the unsettling possibility that others are being paid much more than they are for each patient.

Then there is the issue of recruiting. Here the company and the clinicians have the same ultimate goal: They want to enroll the largest number of patients in the shortest amount of time. The clinical investigators are sometimes spurred on by the promise of a "performance bonus" of perhaps $500 per patient recruited before a certain date. The longer clinical trials last, the more costly they are. From the company's point of view, time is money.

In the past, some dishonest doctors have taken advantage of this corporate eagerness to recruit. A California practitioner, Robert Fiddes, went to prison after it was discovered that he was faking patient records to enroll them in studies. By the time he was arrested, his research institute had conducted 170 studies involving virtually every drug company in the country, and made millions in the process. Fraud on such a grand scale is probably rare. But AutoImmune's Jo Ann Wallace says she has witnessed other examples. "Doctors," she notes, "are as greedy as anyone else. Some guys will try and put their dead mother in a study if they think they can get a buck for it." Not all companies are blameless either: recently, an executive of a company called BioCryst Pharmaceuticals Inc. was sentenced to prison, along with his wife, for falsifying data to win FDA approval of a drug for skin cancer.

"Dry-labbing," as the practice of faking records is sometimes called, is the worst case—a complete breakdown of the alliance between the company and the investigators. But there are honest differences as well. As much as they want recruits for the study, and want them quickly, a legitimate company wants to make their "endpoint" even more. They want the numbers to be strong enough, especially in a critical phase 3 study, so that they can get the drug approved by the FDA. Statistical significance, the company's holy grail, is partly dependent on getting the right patients into the study. The company wants the doctors to play by its rules, which means excluding patients who don't meet the criteria, and sticking to the protocol once they're admitted.

Doctors, for a number of reasons, want to bend the rules. In many cases, the doctor's concern for the patient creates a conflict. A caring doctor like Lea Sewell sees a patient who is discouraged and hasn't been helped much by her interventions. She believes, based on her experience, that the patient is right for the study. But there is one number, or piece of history, that doesn't fit. In the phase 3 Colloral trial, for instance, a woman in her thirties with severe RA who wanted to get off methotrexate seemed a perfect candidate, and was eager to come into the trial. But her sedimentation rate (the marker in the blood that is associated with RA) was just a couple of points below what was required in the protocol. No amount of special pleading for this patient would change the company's mind. As Jo Ann Wallace puts it, "We're not about to screw up the database."

Malcolm Fletcher, who treated patients before he became a trial administrator, explains that "the focus one has to keep in this game is that what I am doing is not trying to help individual patients, what I'm trying to do is to help populations who have a disease. And if you allow yourself to get distracted by trying to help individual patients, it can kill you. The minute a patient enters a clinical trial they become part of something that's more important than they are."

Doctors, however, are imbued with the belief that the welfare of the individual patient transcends all else. In a time when they get little respect from the managed care companies, they want a little from the drug companies. They believe that their professional experience deserves to be taken into consideration by the company executives, who have little or no contact with real people suffering from the disease in question. And they know, from clinical experience, that a lot of drugs work on patients who don't exactly fit the protocol's requirements. The investigator meeting in Charleston played out this physician/company conflict over and over.

The phase 3 trial, Fletcher told his audience, was the twelfth human trial of Colloral. Then he went on to chronicle the ups and downs of AutoImmune's previous Colloral trials, complete with the emotions they evoked—"we got very excited at this point," "we were rather depressed at this point," and ending with an apology because he went through the whole story at "rather a gallop."

Malcolm Fletcher was working, as he always did, to amuse and befriend his listeners. But Jo Ann Wallace, who was sitting in the back of the large conference room, was not pleased. "I have to speak to Malcolm about his presentations," she murmured. Much of what he said had already been explained in site visits before this meeting. There was no need, in her opinion, to go on so long.

Since the Myloral failure, Jo Ann Wallace's imprint on the company had become increasingly visible. Some found her bossy and dictatorial; at the least, she could be called a stickler. She was somebody who didn't read the *Boston Globe* because she couldn't stand what she considered an excess of typos in it, and who thought it would be lovely to live in Singapore, where the rules were enforced. At AutoImmune, she insisted that the key to success in phase 3 was enforcing the protocol. Malcolm Fletcher, with his tendency to see every side of every issue, was definitely not her style. Joe Boccagno, who rose to take questions with Malcolm after the opening presentation, was more to her liking: predictable, good at the details, and perfectly dressed, as a J and J man would be.

Boccagno went straight to the issue at hand: criteria for admission in the phase 3 trial. Right away, there were questions from doctors in the audience. A California doctor with a blonde flip rose to ask about the list of pain medications that patients were allowed to take in the trial. Could certain "analgesics" that were not on AutoImmune's list, and were the only ones paid for by California Medicaid, be added?

"I think the answer is yes," Boccagno responded. But then he realized he may have been too hasty and turned to Fletcher beside him. "I don't know. What do you think, Malcolm?"

"Joe has given you a nice commonsense answer," Fletcher responded. "Let us take this offline and we'll look into it."

This little exchange was a wake-up call. Here was a relatively simple request, to add a couple of analgesics to the list, and yet it was greeted with a dodge: We'll take this "offline," wherever "offline" was. AutoImmune, which had been known to negotiate in the past, wasn't going to budge on anything this time around.

Still the doctors pressed on. A rheumatologist in his sixties rose to suggest that the sedimentation rate required for admission to the trial was high. In his experience, males with RA tended to have lower rates and were unlikely to qualify. Joe Boccagno's answer was polite but firm. "We've pretty much decided to keep it at 28." The doctor's shoulders tensed and narrowed visibly as he sat back down.

Various other doctors were concerned about the fact that the protocol called for patients with "a lot of arthritis." Specifically, AutoImmune was looking for patients with inflammation in ten to twelve joints, and was asking them to go off the strong drugs, like methotrexate, they may have been taking before entering the trial. They would be allowed only pain relievers and a consistent dose of steroids during the study. A white-haired doctor with a ruddy complexion pointed out that patients who joined this study were likely to have flare-ups. What if a patient has a flare-up in the knee that keeps her from walking? Could he give a local injection of a steroid?

Malcolm and Joe agreed that a steroid injection would not be acceptable under the protocol. Patients were allowed to be on steroids and stay in the trial, but the dose had to be kept constant. "Anything we can do to reduce the noise level, we'll do," Fletcher explained. "If steroid perturbations occur, you introduce noise into the study."

"What if I decide to give an injection?" Martha Barnett asked.

Fletcher answered. "If you were the instigator, you would need to be aware that would end the patient's participation in the trial."

A titter went through the audience at Fletcher's use of the word "instigator."

Joe Boccagno, hearing the reaction, turned to Fletcher and commented, "That was very tactfully put," a jibe that got one of the few laughs of the morning.

During the morning break, there was entertainment, along with a second breakfast, in the sunny foyer outside the meeting room. But it looked, at first, as though the entertainment was going to make everyone even more uneasy than they already were. Four black women, dressed in bright plantation costumes complete with colorful bandannas, appeared and burst into a cappella song, beating time with tambourine, gourd, and sticks. The contrast between the luxury hotel setting and the evocation of plantations and slavery made everyone in the nearly all-white crowd a little uncomfortable. Was this the old South or the new, or some weird mixture of the two? Fortunately, the singers were good, and their renditions of "Certainly Lord" and "Dry Bones" soon had everyone smiling and moving their feet.

"The people who do this work," Lea Sewell observed, "are people who pay attention to details." At the Charleston meeting most of the time was devoted to going over, in detail, the procedures required of the investigators. Orysia Komarynskyj, from AutoImmune, discussed the "investigational product," or IP, made from the sternum of chickens, and passed around a little white plastic bottle, with a screw-off top, in which each patient would receive their Colloral. "It should be cool to the touch," she explained, "when the patient takes it out of the refrigerator." However, it was not acceptable to freeze it, as one patient did, then stick it in the microwave to thaw it out.

Malcolm Fletcher, standing up front, demonstrated what the patients must do every morning, dispensing three drops of liquid into a glass, to be filled with orange juice.

Orysia was followed by employees of the CRO (clinical research organization) AutoImmune had hired to conduct the trial. The head of the company, a 250-strong organization called Premier, emphasized communication. "If you're not sure whether or not to bother us," Chris Gallen said, "bother us." Gallen, with rumpled shirt, unruly eyebrows, and potbelly, looked like the last person who should be put in charge of something so obsessive as a clinical trial. But his broad introduction was followed by very precise presentations by the women on his team, who spoke about how to ship blood samples and medicines and how to record data. It was all very necessary, but also very predictable.

The surprise of the afternoon session came when investigators learned that AutoImmune was going to hire a consulting company to recruit patients to the trial. Working with consultants was nothing new: Small companies like AutoImmune relied routinely on CROs like Premier. But Jo Ann Wallace had decided, as a way of getting patients into the study quickly, to undertake a television advertising campaign, with spots on *Oprah*, *Jeopardy!*, and other popular programs, coordinated with phone screening by a company called Pharmaceutical Research Plus (PRP). This got the doctors' backs up.

The same doctor who had asked earlier about the sedimentation rate now wanted to know what would happen if he had a patient in his office who was a good candidate for the trial. Could he enroll the patient right then and there? He was told that he would have to put the patient in touch with the 800 number for PRP.

"Why do that?" the doctor responded. "I've been doing this for twenty-five years. Why defer to someone who's been doing it for only two years?"

The answer, from Malcolm Fletcher, was a variation on the company theme. "It has to do with consistency."

Similarly, the questions from doctors were variations on *their* theme: They were professionals, and their judgment should be trusted. Not only did they object to the idea of turning recruiting over to a phone screener, but they also didn't like the screening questionnaire, which was a blunt instrument compared to the screening they, with their years of experience, could do.

There was, for instance, a question about whether the patient had used any fish oil in the last three months, and another about whether they had taken collagen, cartilage, or "any nutritional products for your arthritis" in the last six months. It was meant, Malcolm Fletcher explained, to get at the "funny remedy" crowd.

But the doctors protested that practically every patient who ever had RA had tried a "funny remedy" at some point. "Have you done a phone survey?" asked a doctor from Washington. "Half of the RA patients are experimenting with home remedies! That exclusion is going to be a killer."

The longest debate concerned the question about "morning stiffness," an important indicator for RA. The questionnaire asked, "Do you have morning stiffness?" and "How long does the stiffness last?" The patient had a choice of two answers: less than an hour, more than an hour. Lea Sewell pointed out that "morning stiffness is a very hard question to get a clear answer about. It's hard to teach residents in training about morning stiffness." You need to ask the question several ways, Sewell insisted. "When does it get as good as it's going to get?" for instance, and "When does it wane?" None of this would be ascertained by the phone screeners, with their simple-minded questionnaire.

The spokesperson for PRP, a bleached blonde in her thirties with bright red lipstick, did her best to respond to the objections from Sewell and others about the phone questionnaire. It was, she pointed out, merely a preliminary step. The doctors would have a chance to talk with candidates before they were enrolled in the trial. But some of the doctors feared that the phone screening would eliminate good candidates and include others whom they would have immediately disqualified.

"You know," Martha Barnett commented afterward, "I think AutoImmune's gotten sort of backed into a corner about this, and they're defensive, which is scary. They have knee-jerk reactions, trying to keep it so perfect, without any real common sense. And they want patients with really severe disease, between ten and twelve joints, I mean that's a *lot* of arthritis. They don't want *anything* to be an outlier, they want everything to be over that hurdle. And beyond the exclusion criteria, once you talk to patients and say they have to stop the drugs they're on, there's a 50 percent chance of being on placebo, and even then they're going to get a drug that *might* work, it's not going to be easy to recruit."

There was a moment during the day when Malcolm Fletcher confided that he was "ready to kill" one of the nitpicking doctors.

But by the time they gathered for dinner at a restaurant carefully selected by gourmet cook and wine connoisseur Jo Ann Wallace, the AutoImmune crowd was in a giddy mood. The conversation turned to the absent Bob Bishop, who was on a sailing trip. All agreed that he would have wanted to go to McDonald's instead of a fine restaurant. "Even in Japan," Joe Boccagno said, "Bob can find the golden arches."

For Jo Ann Wallace, who likes to serve a palate-cleansing sorbet between courses when she has people to dinner, Bishop's taste in restaurants made them incompatible traveling companions. "It got so Bob wouldn't even tell me where he was going for dinner," she says. "He'd say, 'Well, I'm not really hungry,' then slip out later and go to McDonald's."

As they thought back over the day's proceedings, the group agreed that they had held up well to the onslaught. Wallace and Fletcher had known beforehand that the phone screening and the questionnaire would raise objections among the investigators. There had been a big debate about whether even to include the prescreening questions in the handout.

But more important than the phone screening questionnaire was the protocol for the trial itself. On that, Joe Boccagno pointed out, they had held firm. In the phase 2 trials, in his view, "the physicians really weren't instructed firmly enough and frequently enough about how we wanted the clinical studies run. And so, taking that experience, we structured something very different."

"My theory is that they went after the questionnaire," Joe Boccagno concluded, "because they couldn't budge us on the protocol. We have one shot. We need to do it right."

March 26, 1999. An operating room at Brigham and Women's Hospital, 7 A.M.

Howard Weiner is supine on the operating table, his long, muscular body draped, with only his face and his right knee exposed. He is awake, but numb from the waist down, and he is watching on a television monitor as a surgeon invades his right knee to repair the anterior cruciate ligament. At the end of the operation, the surgeon raises Weiner's foot up and rests it on his chest, so that he can get at the knee to wrap it. Weiner tries to wiggle his toes, but nothing happens. The foot is not under his control. It occurs to him, as he lies there, that he now knows what it feels like to be an MS patient with paralyzed legs. It is, Weiner writes in his journal afterward, "frightening to say the least."

Reflexively, Weiner accentuates the positive. The surgery is, he notes, "an important experience for a neurologist who takes care of MS patients." But for a man approaching his mid-fifties, who has been preoccupied for most of his life with mortality, the wounded knee is but the latest reminder that there are limits to what he can achieve. Just a week before he injured it, Weiner had a conversation with a colleague about aging scientists who can't keep up with the latest developments. "We are like professional athletes," he wrote in his journal then, "performance-driven until we are no longer able to compete."

More and more, Weiner found himself wondering how much time he had, and what would happen to him when time ran out. He wrote down a quote from Woody Allen in his diary: "Some people become immortalized by their words . . . I'd like to become immortalized by not dying." The day before the injury, Weiner had spent a happy Thanksgiving with the boys, Mira, and friends, including a special guest, the Israeli novelist Aharon Appelfeld. He tried then to engage Appelfeld on the subject of death, but didn't get very far. Appelfeld explained that he couldn't imagine himself dead, and didn't like to think about problems he couldn't solve.

The next day, while he was playing five-on-five basketball with Dan and Ron at the club, Weiner jumped up to grab a rebound. He got the ball and hung onto it, but he was bumped hard on the way down and landed off balance. Right away, he felt a snap and a sharp, shooting pain. Later in the day, when he tried to play squash, he realized something was seriously wrong with his right knee. The doctors explained to him that he had torn the anterior cruciate ligament (ACL) and that he had a choice: He could live with it, but he wouldn't be able to ski or play squash. Or he could have the surgery, with its concomitant risks, pain, and prolonged recovery time. For Weiner, the choice was obvious. As Churchill had put it: never ever ever ever quit.

It turned out to be the right choice. There were a few days of excruciating pain following the surgery, and they were made worse because he decided to do without Percoset and its side effects. But on the fourth day after the operation, Weiner was back in the MS clinic, seeing a patient who had traveled from South America for a consultation. He was on crutches, and the patient was on crutches. But there was all the difference in the world between their situations. Within weeks, he could throw his crutches away. The MS patient, on the other hand, faced the prospect of moving from her crutches into a wheelchair. Nothing Howard Weiner had to offer was likely to reverse that course.

Despite the failure of the Myloral trial, however, some things had happened to make Howard Weiner feel successful. Not long after he was named a professor, neurologists from Mass General and the Brigham met to decide on a consolidation of the two hospital MS clinics, and Weiner, who had first studied MS as a lowly resident under Barry Arnason at Mass General, was the unanimous choice to lead the combined service. That meant that the new MS center Weiner was planning, and the new MRI magnet, the first in the United States dedicated exclusively to the study of MS, would serve one of the largest populations of MS patients in the country.

Yet Weiner still felt the failure of the Myloral trial acutely, especially when he witnessed the success of other MS drugs. A visit to Biogen, just across the river in Cambridge, in the fall of 1998 served as a reminder of just how high the hopes had been, and how deep the disappointment. Biogen, as he noted in his journal, hadn't even believed in their MS drug, Avonex, in the beginning and had to be persuaded by a committed scientist to pursue trials.

Owing in large part to a kickstart from Avonex, Biogen had become one of the fastest-growing companies in Massachusetts, with 1,200 employees worldwide and 800 in Cambridge alone. "It is somewhat painful," he wrote in his journal, "to see Biogen expanding and doing so well, based on a single drug for which the mechanism of action is not known. At AutoImmune, all was so logical." Of course, he told himself, it was not over yet.

In one way, the visit to Biogen was gratifying. Ahmad Al-Sabbagh, the Lebanese scientist who had started out as a technician in Weiner's lab, was now a Biogen executive, using his language gifts, his medical training, and all the years of experience he gained working on oral tolerance to oversee some of Biogen's clinical trials. His office was large and bright and the windows provided spectacular views across the river to the Hancock building. Ahmad Al-Sabbagh always told Howard Weiner, with a lot of feeling, that he owed all his success to him. Still, seeing Ahmad was bittersweet. It brought back memories of the early days, when Howard and Ahmad had worked together, in excited anticipation, at the messy business of isolating myelin from cow brain for use in the very first human trials of oral tolerance.

At AutoImmune, Bob Bishop was upbeat. He was convinced, as he told investors at Hambrecht and Quist early in 1998, that AutoImmune was "back on track" and would succeed with its phase 3 trial of Colloral for rheumatoid arthritis. Howard Weiner was hopeful too. But as much as he hoped Colloral would succeed, and bring relief to arthritis patients, Weiner's deepest commitment was to MS. MS was the disease he had vowed to find a cure for, thirty years earlier. And he was impatient about getting on with it. Ever since the Myloral trial failed, he'd been casting about for a way to try oral tolerance again with MS patients.

As it turned out, the best opportunity to try again came from research that had resulted in the C drug of the ABC drugs for MS, Copaxone. Teva Pharmaceuticals, the Israeli-based company that produced the drug, might have been viewed by some in the biotech business as the competition. But Weiner had maintained active and cooperative relationships with other companies working on MS. Because of his Israeli affinities, the ties to Teva were particularly strong.

The Copaxone story began in Israel in 1967, when Weiner was still a medical student in Colorado. Scientists at the Weizman Institute were looking for protein fragments that would induce EAE in animals, and they put together four different amino acids, based on the sequence in myelin basic protein, into a compound they called Cop 1. To their surprise, far from inducing EAE in animals, Cop 1 seemed to protect the animals against the disease.

It took a long time for another researcher, Murray Bornstein at Albert Einstein in New York, to take up the work begun at the Weizman Institute and attempt to apply it to human disease. But in 1987, the *New England Journal of Medicine* published promising results of a human trial of Cop 1. Howard Weiner, who by then had completed his first successful oral tolerance experiments in animals, wrote an editorial for the *New England Journal of Medicine* in support of the Cop 1 trials. "The results demonstrate a beneficial effect," he wrote, "in a subgroup of patients studied—those with little or no disability at the beginning of the study." In the editorial, Weiner also raised the question that would plague believers in Cop 1 from that time on. "If Cop 1 really worked, how did it work?"

Cop 1, it was sometimes suggested, was a discovery made by researchers who had "more luck than brains." It worked by an unknown mechanism on a disease whose mechanism was poorly understood. That may have been one of the reasons it took so long for Cop 1 to find a sponsor in the profit sector. But in the early nineties, Teva Pharmaceuticals undertook a large, two-year, multicenter phase 3 trial and came up with statistically significant results. Cop 1 was no miracle drug, but it had a measurable effect on the attack rate in early-stage MS patients, reducing it by approximately 30 percent. In September 1996, Howard Weiner attended the FDA hearing where Cop 1 was approved for the treatment of MS. As a regulator noted at the hearing, one always hoped for the next penicillin. But in the meantime, anything that worked was better than nothing at all.

One member of the FDA panel brought up the old issue: "No one could quite say," he noted, "why it does what it does." Howard Weiner was one of those who suspected that Cop 1 might be operating by a mechanism similar to that produced by the whole protein, myelin basic protein, in oral tolerance. If that were the case, then perhaps Cop 1 could be given orally, rather than injected every day. This was an idea that appealed to Teva Pharmaceuticals enormously. All the ABC drugs had to be given by injection, and that meant a lot of patients were resistant to taking them. An oral drug that was equally effective would drive competitors out of the market in no time.

By February 1998, Howard Weiner had Ruthie Maron, the self-avowed "rabbit," doing feeding experiments with Cop 1 in mice. Both TEVA and researchers at the Weizman Institute were doing similar experiments. In March 1998, he and Bob Bishop traveled to Israel to negotiate with Teva about patent issues related to their plan to conduct human trials of oral Cop 1. Scientists representing Teva had suggested that Cop 1 worked as a result of its cross-reaction with myelin basic protein and induced "bystander suppression." That meant that an oral trial of Copaxone, if it took place, would be based on ideas that had been developed by Weiner and patented by AutoImmune. Many months of negotiation followed the first meeting between AutoImmune and Teva in Israel. But finally, a year later, an agreement was signed. Teva would undertake a human trial of oral Copaxone. And if MS patients benefited, AutoImmune would, through its patents, reap some of the monetary rewards. It was a triumph for AutoImmune's patent lawyers. But far more important, as far as Howard Weiner was concerned, it was a second chance at a major trial of oral tolerance as a treatment for MS.

In April 1998, two months after the agreement between Teva and AutoImmune was signed, Howard Weiner presented his lab's experimental work on feeding Cop 1 to mice at the annual meeting put on by FASEB (Federation of American Societies for Experimental Biology) in Washington, D.C. Actually, it would have been more appropriate for Ruthie Maron to present. She had done the work, and had already given a preliminary paper on it months before. But she was always reluctant to get up in front of a crowd, explaining, "Howard has what I lack—confidence." So late on the night before the presentation, Anthony Slavin delivered the slides to Howard Weiner, who was staying at the Marriott across from the convention center, and it was he, not Ruthie, who got up on the platform the next day to talk about oral tolerance and Copaxone.

About 12,000 scientists had come from all over the world to attend the annual FASEB meeting. The program for the meeting was the size, as Anthony Slavin noted, of *War and Peace*. There were two volumes of abstracts besides, each about the size of the Boston phone book. To an outsider, seeing the hundreds of people waiting in long lines for credentials at the cavernous Washington Convention Center, it seemed unlikely that anything worthwhile could be gleaned from a meeting on such a scale. But there were, of course, many hundreds of small neighborhoods within this city. One of them, the one where Weiner and his group dwelt, was the growing area of immunology called oral tolerance.

At 10:15 on Saturday morning, the five members of the Weiner group who had flown down from Boston for the meeting were sitting, shoulder to shoulder, in the front row for a panel on "mechanisms of oral tolerance," chaired by the oral tolerance pioneer from Ohio, Caroline Whitacre. The room, which could hold about 500, was two-thirds full for the usual cluster of eight fifteen-minute presentations. For the most part, the presenters were postdocs, many with accents. Weiner, the last to speak, was by far the most senior.

It was the first time Howard Weiner had traveled since his knee surgery. At the airport, he'd used a wheelchair. When he walked, he moved slowly, and asked other people to carry his briefcase for him. Even though the Marriott was a block away, he took cabs between the convention center and the hotel. The knee didn't stop him, though, from rising and walking to the audience mike in the aisle, after four of the eight presentations preceding his, to question or comment. He was, as his postdocs noted with amusement, extremely good at asking a question or making a comment that included some reference to the work going on in "our lab." Also, even though he hadn't had much time to master the slides, Weiner presented Ruthie's work almost flawlessly. The only time he stumbled, saying IL-6 instead of IL-10 because he couldn't see the slide well from where he was standing, his postdocs in the front row quickly corrected him, in unison.

Toward the end, when he put up the most colorful slide of the talk, Anthony and Ana Faria, the new Brazilian in the lab, exchanged satisfied grins in the front row. Using circles, diamonds, rectangles, and blob shapes in bright colors, the slide showed the action of Cop 1 via the injected route and via the oral route, and suggested the mechanism by which the compound protected against EAE in mice. There had been a lot of discussion back at the lab about which shapes and colors to use in this slide. Should Cop 1 be a triangle or a diamond, and should myelin basic protein be a different shape, but somehow related? What should be red, what yellow, what blue, in this abstract condensation of the biological action under discussion? It was, Weiner said afterward, a useful process: while they were debating how the slide should *look,* they were also working on the hypothesis, and on understanding the mechanism.

At a meeting of neurologists, everyone would have been familiar with Copaxone, because they were using it to treat their MS patients. But the Washington meeting was a gathering of basic scientists, and they were unfamiliar for the most part with this rather odd concoction, which had been stumbled on by chance and which seemed, according to Weiner's presentation, to work even better than myelin basic protein in the mouse equivalent of MS. Chuck Elson, a senior and respected immunologist at the University of Alabama, rose to ask the familiar question. How could this Cop 1, this cluster of protein fragments, work? The answer, as Weiner explained again, was that it triggered "bystander suppression"—it activated T cells that suppressed a broad spectrum of inflammatory agents.

As the hall emptied, the little group from the Weiner lab clustered to evaluate the presentation. Ruthie declared that Howard did a "wonderful" job presenting her material. But they all expressed frustration that Chuck Elson, one of the leaders in the field, didn't seem to entirely accept that Cop 1 was operating by "bystander suppression." "It's funny," Weiner observed, "but sometimes when a scientist hears something for the first time he doesn't believe it, no matter what the data." Sometimes, Ruthie noted, it's a question of where it's coming from. "Some people say, 'I don't trust anything that comes out of that lab.'" But sometimes the labs that are suspect are doing the most interesting work.

As everyone knows who has ever attended a professional conference, the formal presentations are usually the least useful part of the experience. It is what goes on in the interstices—the informal remarks after the talk and the chatting over lunch and dinner, that are most instructive. Then, too, there are the comments that often come at the end of a paper, about work that's too recent to have been in the abstract. That's the part, Ana Faria says, that she always listens for.

Scientific meetings like FASEB also have poster sessions, where researchers put up a schematic of their work on a large board and stand by it, elaborating for colleagues who stop by. Poster sessions provide a chance to talk informally about the work and to make connections, provided you know the faces.

Everybody still teases one of the Japanese researchers in Weiner's lab, Yoshi Komogata, about the time he didn't. Yoshi was very interested in working at the Center for Neurological Diseases. So when he saw Weiner at a meeting, he talked with him about the possibility of a job. But later that same day, when Weiner stopped by to see his poster, Yoshi didn't recognize him. Yoshi, whose Japanese upbringing makes him especially respectful of his superiors, was caught up in a conversation with some other people and sent Weiner away with barely a glance. After he joined the lab, the other postdocs liked to remind Yoshi of the time he snubbed Howard Weiner.

At the FASEB convention in Washington, Howard Weiner limped slowly through the lines of posters on oral tolerance. Since his lab was presenting a total of fourteen abstracts and posters at the meeting, some of the displays were very familiar. Even the poster-tenders he didn't know brightened when he passed by. One woman, who had been experimenting with a new oral approach to MS using interferon-tau, practically fell upon him. "I want to meet you," she said, "and say thank you because the *only* reason I've gotten any support for work with interferon-tau is that Howard Weiner's interested in working on it."

Halfway through his poster tour, Weiner ran into Caroline Whitacre, who was making a similar visit, and the two of them talked about a poster that was at the center of a controversy at the moment, having to do with the anomalous behavior of a certain strain of mice. Whitacre stood with her arms folded in front of her. Several postdocs had gathered round to hear what the two senior scientists were talking about. "What does the boss think?" Whitacre asked Weiner, then laughed at herself for saying it. There was a hint of sarcasm in the question, since both she and Weiner were bosses. But neither of them was about to show anything but respect in this public setting, whatever resentments might be simmering beneath the surface.

After his tour of the posters, Weiner took a cab back to the Marriott to prepare for the next stop: He was going directly from D.C. to the American Academy of Neurology meetings in Montreal. He had packed a larger bag than usual for this two-part junket. It was always his policy, as he explained, to "take a little more than I need." But this time he also packed his computer and as many papers as possible into another large bag, so he had less to carry on. As he folded in one pair of pants after another, he talked about the future of the Cop 1 research.

"I think it's a very interesting next step. And it actually gets into some of the new theory we're working on, in terms of altering the structure of the antigen. So I think it's a natural. And I'm very excited about it. I think it has a very good chance of working."

Even though he would like to have seen AutoImmune succeed with Myloral, Weiner insists that a successful trial of oral Copaxone would be "very satisfying."

"Nothing you do in science," he points out, "is only yours. They did the oral Copaxone because of everything I did. They're going to do a big trial of something like 1,200 patients. That could turn out to be oral tolerance for MS!"

In March 1999, twelve years after his first editorial on the promise of Copaxone for treating MS appeared in the *New England Journal of Medicine,* Howard Weiner wrote another editorial about oral Copaxone, this time for the PNAS. "It takes as long as 15–30 years," he wrote, "from the time of a basic science observation for that finding to be reduced to practice in the clinic. This time frame has been true for Cop 1, which was first described in 1970, and also may be true for the optimal application of oral tolerance to human disease."

CHAPTER 15

Human Trials

August 1999, Swansea, Massachusetts

Manuel Barboza, a retired factory worker, is standing in his driveway, demonstrating the amazing effects of Colloral on his debilitating rheumatoid arthritis. "Before," he says, "I used to walk like this." He advances in little mincing steps, scuffing up the gravel with his black sneakers. "But now I walk like this." He strides forward, with back straight and arms swinging, feet lifted high, as though he were the drum major leading a marching band. The Colloral, Barboza insists, is "a miracle drug. It just sneaks into your body. Once people start taking it, they'll be shocked."

Manuel "Manny" Barboza, a black man with curly salt-and-pepper hair and a bald spot on top, likes to charm people. Wherever he goes, he carries hard candies in his pockets and hands them out. At the sandwich shop, he gets a laugh out of the teenage cashier by asking if he gets a discount for being "good lookin'." He is, in fact, a good-looking sixty-five-year-old with big, strong hands and taut limbs, the product of a life of physical labor that began at thirteen, when he started sanding down cars after school at the body shop behind his mother's house in Fall River.

There were factory jobs—at a leather company in Taunton, at a rubber company in Stoughton, and finally at J and J Corrugated Box Company, where he worked for twenty-eight years before his recent retirement. But all the time he was doing these jobs, he was working at other jobs too—drove a wrecker for a few years in his off hours, worked for an Italian who built cement chimneys, and worked for a time too as a cook at the Clearwater Lounge. He even held the job of Swansea dogcatcher. His best second job was the most physically taxing of all: working as a stevedore unloading the big boats that come into the harbor at Fall River. Sometimes it was bananas, other times it was ninety-two-pound bags of cement—"all day long, forty bags on a pallet."

Barboza's kept up his membership in the Longshoreman's union, and he's eager to get back to working on the docks, now that he's feeling better. It's good money, he notes—$17 an hour and even more in overtime. "Hey," Barboza says good-naturedly, "I ain't afraid of work."

Manuel Barboza's parents came from the Cape Verde Islands, seeking work in the mills that made Fall River, at one time, a great center of textile manufacturing. He married at twenty-one, to a woman who also had Cape Verdean roots, and they bought a house "in the country" together, north of Fall River, where they still live. His wife, like him, is retired, but still works several days a week at the HiLo grocery store.

On this summer day, Barboza sits at a weathered plywood table in his yard, with the radio turned on low, surrounded by the comings and goings of his family. One of his five children, a beautiful daughter in a chic little black dress with a big head of black hair pulled into a loose ponytail, emerges from the house. "That's her style," he says with a twinkle. She is off to her job at Lord and Taylor. A tetherball hangs from a nearby tree, next to a plastic slide, and his two grandsons are circling nearby. "They're a nuisance most of the time," he says, but he's careful to notice whether they're in the cab of the pickup when his son goes off on an errand. "I only see one in there," he says, "the other one must be in the kitchen with my wife." His son and daughter-in-law live a few yards away, in a neatly kept double-size trailer. The son's business, an auto body shop, occupies a multicar garage on the same property. There is a twenty-seven-foot trailer for trips, which Barboza plans to make use of once again this summer. "Last year I couldn't go nowhere," he says. "Now come next year, I'll be able to run alongside the mobile home." Barboza is dressed for a summer day in shorts and a T-shirt, which reveals a modest belly. That will go, he says, patting it, once he can get back to work.

Given his work history over a half century, it wouldn't be surprising if Manny Barboza had a few aches and pains. Yet it wasn't until he was close to retirement from the J and J box company that he was laid low by crippling rheumatoid arthritis. His ankles and wrists swelled to twice their size, and his feet were so inflamed and swollen that he felt like he was "walking on a sponge." The simplest things became impossible. "My wife was tying my shoes for me, and I couldn't put my stockings on, it hurt so damn much."

Barboza demonstrates what he had to do to wash up: He lifts up one arm, stiff at the elbow, with the other and props it up against an imaginary wall, then washes it, brings it down, and repeats the same routine on the other side. Sometimes, even moving his thumb a little hurt so much he had to grab it and wince—"you'd get that pain like a wicked charley horse." The things he used to do at work, like moving around five-gallon buckets of ink, became impossible. "Some days," he says, "I woke up in so much pain, I wondered, should I go to work or should I go to the hospital?" But he always went to work, determined to hang in there until his retirement date.

He also went to see a Fall River rheumatologist named Ronald Rapoport to find out if there was anything to be done. Rapoport put him on Lodine, an anti-inflammatory drug. When that didn't seem to help, he told him about AutoImmune's Colloral study. The way Barboza tells it, Rapoport pretty much made the decision for him. "He says, 'I'm gonna put you on a study. A new study.' He says 'I signed you up already.' So I went on it for six months. And everything just started getting back into shape."

Barboza felt good enough, after he got in the trial of chicken-derived Colloral, to kid around about it. When Dr. Rapoport and his research coordinator, Arlene Turgeon, asked him on one visit about side effects, he told them there were some.

"What's the problem?" Dr. Rapoport wanted to know.

"Well," says Manny, "I caught myself in the morning going like this." He flaps his elbows out to the sides, then flaps his hands, doing the movements of the chicken dance.

"They cracked up."

What made it more than funny was the fact that, only weeks before, Manny Barboza hadn't even been able to bend his arms at the elbow. The three of them walked out of the office, according to Barboza, moving to the beat of the chicken dance song.

It could have happened. The Rapoport office can be informal in a way that would be rare at a Harvard teaching hospital. "Dr. Rapoport is a character," explains research coordinator Turgeon. "If he has a day where he's giddy, nobody gets anything done." Turgeon herself, a high school graduate who has learned how to run research trials over the past thirteen years, is the most medical-looking person in the office. She wears a crisp three-piece white uniform that sets her tan and her maroon nails off nicely. A pencil, stuck in her thick, long hair, makes her look as though she means business. And her small office is no-nonsense: One wall is taken up with locked metal cabinets and a locked refrigerator, in which study drugs for various trials are neatly stacked. In the seventy-plus trials the Rapoport office has been part of, Arlene, as she is universally known around the office, is the one who has stayed on top of the paperwork and kept track of every ounce of drug. She is proud of the office's good reputation as a clinical trial site.

Her boss, across the way, has demedicalized his office. Except for the eleven degrees and licenses he has hanging over his desk, Ron Rapoport could be your lawyer or your accountant. He sits in a desk chair that swivels toward the patients when they arrive, and the patients face him in two office chairs. If necessary, they repair to the room next door, where there is an examining table. But for the most part that's only when he's going to give a shot. Otherwise, it's mostly conversation, and usually not a lot of that.

A compact, fit man with deepset eyes and a vertical crease, like an exclamation mark, above his nose, Ron Rapoport wears an open white shirt and chinos to work. In his early fifties, he is, in almost every way, the antithesis of an academic doctor like Lea Sewell, who listens carefully to the history, does the complete careful exam, then talks through the alternatives. Instead, he listens for the key complaint, then advises, in no uncertain terms. There is much less talk, more action. Shots, sample drugs, and prescriptions are given out more often than not, along with a lot of kidding around. His appointments are brief, because there are a lot of them to get through in a day. On an average Monday, Rapoport saw an astonishing forty-three patients.

Early in the day, a small bent-over woman in shorts and a T-shirt comes in to complain about her knee. She's tried Vioxx and Celebrex already and they don't seem to help. But her chiropractor suggested gelatin and she's also heard something about primrose. Before she can even finish, Rapoport purses his lips and makes breathy sounds, then puts his fingers to two sides of his mouth, and whistles. "Gloria, Gloria. Let me tell you. No, no, no, no. They haven't been proven." So much for respecting the patients' need to explore alternatives.

Rapoport is equally opinionated on the other question that comes up over and over in rheumatology. What about surgery? "Maybe you should get a new knee," he says to the woman who wants to try primrose and gelatin, "but only when the pain is such that it's the centerpiece of your day." His answer to the surgery question varies, but he always states his position in straightforward lay terms. You should have surgery, he says, when your knee problem changes from "acceptably unacceptable" to "unacceptably unacceptable."

A short, round woman with a lot of curly white hair who is dressed entirely in purple makes her way in slowly to a chair, depending heavily on her cane. She smiles a lot, but she says she's aching all over, and taking a long time to get mobilized in the morning.

"You have a lot of arthritis in your knees," Rapoport says, "but you're a good girl, you've managed to avoid surgery."

"I don't want surgery," she answers.

"Look at surgery as an option," he says. "If you have money in the bank to buy a new car, you could drive the old car until it breaks down and you need the new one. Then you buy it. That's what the surgery is. You could have that on your knee eventually."

But with another patient, a six-footer in his fifties with severe osteoarthritis in one knee, he takes a firmer approach. "Are you going to have a total knee?" he asks.

Then, "Wait a minute, watch my head." Rapoport nods his head up and down, indicating yes. "I would have done it yesterday."

The patient, whose name is Sean, explains that he's nervous about going under the knife because of a terrifying experience he had with back surgery. He was so close to death after one operation that "They gave me last rites!"

Rapoport listens and says he understands, but he holds to his opinion. "I've seen people with disaster knees come back with smiles on their faces," he says.

Delicacy and discretion don't play a big role in Rapoport's approach. When he decides to dictate a note on a patient, he usually does it with the patient sitting right there. And he doesn't spend a lot of time feeling the patients' pain either. The little woman in purple who's been hurting tells him, as she's leaving his office, that the knee pain is the worst when she pivots. "Don't pivot," he says.

For the most part, though, Rapoport's style seems to work with his patients, who can usually give as good as they get from him. Sometimes, the humor is insulting on both sides. An obese woman boasts to Rapoport about the twenty-two pounds she's lost. Her clothes are getting bigger, she tells him, pulling the elastic band on her pink shorts way out in front of her to demonstrate.

She's come for a Lidocaine shot.

"Where do you want this?" Rapoport asks.

"In my ass!" she answers. "My ass is killing me!"

Rapoport says he tries every time to hurt her, but he never succeeds. "Sometimes I bend the needle a little before I put it in. Sometimes I sneeze and put it in."

Under her breath, but loud enough for Rapaport to hear her, she mutters, "nasty little Jewboy."

Rapoport is unfazed.

"Do you realize it's been twenty years I've been coming here?" she continues.

"That's scary," he says.

"It's scary to think that you could be a doctor that long!" she counters.

The Truesdale clinic, where Rapoport has his offices, is a low-lying brick building on President Avenue, just a few blocks down the hill from the few grand Victorian houses in Fall River. The immediate neighborhood is more typical—rows of modest single homes with porches, some with Madonna statues in the yards, others with signs for upcoming elections promoting candidates with Portuguese names like Oliveira and Silva. The clinic, which is having its eighty-fifth anniversary, looks worn out, like the town. The parking lot is full of potholes, one elevator doesn't work, and it has been a long time since the last renovation.

There was a time when things were better for people living in Fall River, when the mills were running and immigrants from the Azores came and found a better life. Now most of the stately granite mills have been shut down, the windows boarded up, and the huge rooms inside turned into discount stores. Downtown, there are sunny orange banners hanging on graceful swan-neck lampposts that proclaim "Fall River" and "Summer in the City." But many of the shops along the main street are vacant, and the merchants who remain, with the exception of the Portuguese baker, are selling dreams rather than the necessities of daily life. There are at least three bridal shops on the main street, and shops selling christening and confirmation outfits. Machismo is for sale too: there are schools for Tae Kwon Do, Jujitsu, and Kardio Kickboxing, and there is a store where you can order boom boxes for your car. There is also a travel agent offering special prices on airline tickets back to the Azores.

The patients who come to see Ron Rapoport reflect the history and the hard times of the city. A short, dark man in his forties, using a cane, comes in to have Rapoport look at his shoulder. His troubles began when he was hit in the head with a pipe by a member of the mob. Now, he tells Rapoport, he's having trouble with his "ticker."

A lot of the people Rapoport sees have injuries related to dangers at work. A new patient had a freight elevator gate come down on him. He wound up having sixteen stitches on his face and lots of aches and pains all over his body, as well as weakness in one arm. "I can't even throw a baseball anymore," he tells Rapoport, but explains that he went right back to work after the injury because he needed the $200 a week.

Often, the work itself is the problem, but the patients can't afford to stop. One of his patients spends her workdays, beginning at 4:30 A.M., sliding fabric under the needle of a sewing machine. She has weakness and pain, particularly in her left arm. Rapoport tells her, "it's not going to get better until you quit." But she insists that she can't quit. She's worked at this job for forty-two years, and has five to go before retirement. A Portuguese woman who cleans for a hotel has pains in her shoulders and elbows. "Don't kill yourself," Rapoport tells her, peering out over his wire-rim glasses, "unless they're going to give you part of the hotel." But for someone who needs to keep the job, it's difficult to heed such advice.

Many in Rapoport's patient population are among those Americans, now well over 40 million, who don't have insurance. Among those who do have insurance, or are on Medicare, problems arise when they need drugs. "The medicines are outrageous," Arlene Turgeon notes. "Some people have $500 a year for drugs, but if you have a prescription that costs you $100 a month how long is that going to last? People who are older and on Medicare, if they don't have a good secondary insurance, really suffer. It's horrible. People remain sick because they can't afford the drugs."

One makeshift and temporary solution is to give out samples. Drug companies want doctors to dispense samples to patients who can afford to buy them if they turn out to be effective. But when a nurse's aide with significant rheumatoid arthritis tells Rapoport she's had no luck with methotrexate, he goes in the other room and comes back with free samples of Enbrel to keep her going for a while. Others get samples of other drugs. Still, the supply of samples is limited. A patient without insurance or the means to pay for drugs will eventually be deprived of them.

Another way to get free treatment is to enter a clinical trial. "The studies help people," Arlene says. "We have a lot of people who call me and say, 'If you're going to do a study, would you let me know if I would fit in?' We have some people that continually do studies. They're getting medications, they're getting doctor's visits, they're getting all the testing—labs, EKGs, X rays, everything."

But clinical trials are not treatment, as Malcolm Fletcher and others point out. There is a good possibility that the patient will be receiving placebo or a test drug that is ineffective. In rare cases, the test drug might have damaging side effects. And eventually, the study is over. Arlene insists that "professional patients," as they are sometimes known, aren't left without care in the Rapoport office. "We try to put them into a long-term study, something maybe two or three years long, and hope that in the meantime something is going to happen so they'll have health insurance."

"And if they don't," she adds, "believe me, we reduce the rate."

Some believe that it is unethical to substitute clinical trials for genuine medical treatment. "It becomes a further exploitation of the poor," says John Paris, an ethics professor at Boston College. But others argue that a trial is better than no treatment at all, which is what many people are likely to get under the present system. Even if the placebo or drug is ineffective, the patient gets a full workup, as well as the attention of a physician.

As beneficial as they can sometimes be to poor patients, there is no doubt that clinical trials are also sustaining to Ron Rapoport and his operation. "The income from studies is valuable," he acknowledges. Rapoport has an office staff of five, including his wife Jenny, a former surgical nurse who comes in three days a week. And he has three children, ages twelve, sixteen, and nineteen, who go to private school. "That's another thing the studies have helped," he says. "I can afford to send them to private school. And if I did not do studies or give lectures, it might not be affordable."

Rapoport, who checks with his broker between patients and drives a hard bargain with a company that wants to lease him a photocopier, is also attracted to the entrepreneurial side of trial work. He has a card that promotes his clinical research: Phase III is printed in large green letters over a gingko leaf, and underneath, in smaller type, there are five areas of research listed: arthritis, osteoporosis, pain, women's health, metabolic disease. "It's not just making the money that's exciting," he explains, "it's making a successful venture."

In addition to conducting clinical trials, Rapoport is paid by drug companies to give talks to physicians about their product. He likes the special attention the companies give him when he travels for them. Even with patients, he is always pleased by surprise extras. One lady he sees who is a collector of bears brings him some kind of bear every time she visits. He has one on his shelf with wire-rim glasses halfway down its nose that looks a lot like him. And he is excited when she produces a Beanie Baby bear on another visit. Even small things, like little acid-free yellow tomatoes from someone's garden, make him happy in a childlike way.

When the Merck salesman stops by, Ron raves about the drug rep he traveled with on a recent tour of Kansas, where he went to talk with doctors about Vioxx. This particular drug rep gave out gift certificates and Kansas Jayhawk T-shirts for the kids. "She made the doctors feel valued, showed them a good time," he tells the salesman. "That's the way it should be." For whatever reason—because he practices in a depressed city, because he's on his own without collegial support—good treatment from the drug company means a lot. As he comes to the end of his long Monday, Ron Rapoport is on the phone planning for a weekend trip, sponsored by Merck, to a meeting at Loon Mountain in New Hampshire. "I need to be spoiled," he tells the person from Merck on the other end of the line.

Rapoport insists, however, that he isn't influenced by his relationship with the drug companies or by the money the trials bring in when it comes to his patients. "I think I'm a good salesman when I believe my patient could use it. But I don't want to push a study on a patient."

"I'll give you an example. We have somebody in a study now of Vioxx, an anti-inflammatory in RA. It's a one-and-a-half-year study. He's been in it for seventeen weeks. He doesn't feel like he's doing well enough. All I have to do is say, 'Why don't you try it for two more months and see what happens?' The guy likes me, he would do it in a second. I would get paid for two more visits. But I told him to drop out because it was ineffective. It happened this morning. That's the way I try to do this."

"I'm not any more moral than anybody else," Rapoport says. "But I try very hard to be clean when I do studies. For my patients."

AutoImmune enrolled the 772nd and last patient in the phase 3 Colloral trial in February 1999. But long before the last patient was in, various problems at trial sites around the country sent Joe Boccagno and Malcolm Fletcher into action. One experienced investigator had a coordinator who seemed to be an alcoholic, who wasn't following up referrals sent by the screeners. The company had to present an ultimatum: find a new coordinator or lose the study.

Another investigator took a moral position that six months was too long a time for rheumatoid arthritis patients to be deprived of disease-modifying drugs (DMARDs). He argued that the study was unethical, and began talking to other investigators about his view. Malcolm Fletcher had to point out that "at this point in the exercise, you can't change anything. I mean it would just wreck it. Even indulging in this debate would be a serious waste of time." It reminded Malcolm of a movie he'd seen where the bride-to-be backs out on the day of the wedding. She tells her fiancé that she's just decided he's not the man she wants to spend the rest of her life with. The fiancé says, "Well, this is a fact that could have been brought to my attention *yesterday*."

Malcolm, who never had an easy time with confrontation, was especially troubled because this investigator "comes across as a very down-to-earth, pleasant fellow, and I like him a lot, always have liked him a lot." As usual, Jo Ann Wallace was less charitable, suspecting that the investigator was sabotaging the study on purpose, because of his connection with another drug company. Jo Ann told Malcolm, "Get rid of him, and don't waste a second." Reluctantly, Malcolm complied.

Problems didn't always come from the obvious places. An independent operation like Ron Rapoport's, run by a high school graduate who had learned on the job, might seem a likely place for mistakes to be made. But in fact, the error that sent AutoImmune into crisis mode was made at a prestigious teaching hospital on the West Coast. A pharmacist who spoke little English mixed up the drugs for two different studies: Colloral was dispensed to patients in an ophthalmic study by mistake.

AutoImmune picked up the error from looking at their data on the quantity of Colloral dispensed, and reported it immediately to the FDA. To demonstrate their alarm, J. D. Bernardi and the head of quality control for the consultant running the Colloral study got on a plane right away and visited the site. Jo Ann Wallace was more than usually indignant. "They're just morons," she concluded. "The study coordinator had an IQ of less than 80 on a good day. They mixed up clinical supplies from two different companies. It's inexcusable. And the person that actually was probably the number one guilty party was the pharmacist, a Chinese intern pharmacist whose English skills were not very good. It was a situation in a big teaching hospital in which the checks and balances completely broke down." Fortunately, no patient suffered ill effects, since Colloral is harmless. But for AutoImmune, any flaw in the conduct of the trial was a worry. These were the sorts of things that would be noticed when they went to "the Agency" for approval of the drug.

The challenge was, first and foremost, to keep all the sites in compliance with the protocol. But it was also important to try to keep patients in the study. The statisticians predicted that as many as 35 percent of patients would drop out before the study was over, usually because their arthritis was worsening. But Boccagno and others at AutoImmune urged doctors to try to make their patients comfortable enough, using analgesics, to keep them in the study for at least eight weeks, if not a little longer, to give the Colloral a chance to start working. Of course half the patients were on placebo. But the greater the number of patients who stayed in the study, and the longer the length of time they stayed, the more likely it was that there would be a discernible and statistically significant difference between the placebo and Colloral arms of the study.

It helped, as everyone at AutoImmune knew well, if the investigators conducting the study were happy, or at the very least not unhappy. But this, too, presented a challenge, since temperaments and sites varied dramatically. Ron Rapoport, for instance, had no complaints about the phone screening and referral process AutoImmune had set up. "I love it when they screen for us," he said. In general, he was pleased with AutoImmune and with the people he dealt with from the consulting company. His only complaint had to do with the "annoying persistance" of one AutoImmune employee, who bothered him too much about details.

Lea Sewell, on the other hand, was furious about the way the study was being conducted. Her particular beef, as she had anticipated in Charleston, was with the questionnaire phone screeners were using to recruit patients. "They have committed a classic 'no-no.' Rule number one in clinical trials is, if you have a questionnaire, you have to test it. No matter how smart you are you have to try it out because you'll discover things with real people and real patients."

What annoyed Sewell most was that AutoImmune was "not paying a lot of money" and was taking up a huge amount of both her and Michele Finnell's time. They had seen thirty-four people, and enrolled only eight. Michele had spent a lot of time on the phone, even before people came in, doing a second screening to see if the referrals met the criteria for the study. "We've probably made hundreds of calls."

According to Sewell, everyone knows that many people who respond to television ads are poor candidates. "We've had lots of people who call after big news splashes. The vast majority are not eligible, and it's extremely time-consuming, many of them are crazies. People you work with, or people another rheumatologist has worked with regularly are much more reliable. And everybody knows that. And yet Malcolm really believed that people who answer ads would be good. He believed that! He said it in South Carolina."

The company, in an attempt to assuage Lea Sewell, provided her with money to hire a part-time worker to help Michele with the paperwork. But they also defended their methods. The phone screening, according to Joe Boccagno, did bring in a lot of qualified candidates—about 45 percent of the 772 patients in the trial were recruited that way, with 55 percent coming from the rheumatologists' practices or from professional referrals. Without the phone screeners, according to Boccagno, Sewell "would have had about five times the work." Of the 600 people who responded to the television ad, 450 were eliminated by the phone screeners.

Besides problems at the forty-four sites in the Colloral study, AutoImmune had internal issues to deal with. Malcolm Fletcher had been, for personal reasons, itching to move back to North Carolina ever since the Myloral trial failed. His wife had already moved south to be near her children, and Malcolm was toughing it out in a small flat in Lexington. "Living in a room in Massachusetts all by yourself," he noted, "is not wonderful." He bragged, rather pathetically, that he'd discovered a way to make shepherd's pie using frozen mashed potatoes that was almost as good as "scratch made."

Domesticity was on Malcolm Fletcher's mind. Whenever he had a chance, he would furtively pull out the drawings for a house he was building in North Carolina. It was a model called the Churchill, and he was inordinately pleased about its luxurious foyer and gracious layout, and the way it could accommodate all the children, both his and his second wife's. The truth was that, even though he claimed to be "more excited about this technology now than I've ever been," Malcolm was on his way out. "You could discover the supernova," he notes, "but you still have to drive the kid to the basketball game. Which is more important?"

The departure of the clinical director in the midst of a trial doesn't look good to investors. For that reason, Malcolm's departure was kept very quiet for as long as possible. Even after Joe Boccagno had taken over entirely, Malcolm was kept on a small retainer, and checked in periodically with investigators. "When the Street finds out," Jo Ann explained, "they're gonna flip. A resignation of a director of research in mid–phase 3 trial. They don't know the nuances of the situation and they just read trouble. They don't know that Malcolm's wife told him to move to North Carolina or else. We're walking on eggshells over this."

Jo Ann's plan was to wait until AutoImmune had signed the agreement with Teva, in connection with the Copaxone trial, and then to "weasel in the fact that Malcolm resigned."

There were other worries as well. Eli Lilly, which had entered into a partnership with AutoImmune to conduct studies of oral tolerance as a treatment for diabetes, chose not to make its milestone payment. That meant, essentially, that the partnership was dissolved, although Lilly promised to bring the diabetes trials to a conclusion. It also meant a sizable loss of research funds for the Center for Neurological Diseases.

What's more, 1998 and 1999 proved to be banner years for the approval of new rheumatoid arthritis drugs. There were a new class of drugs, sometimes described as "super aspirins," that were easier for the stomach to tolerate: Celebrex was the first to be approved, and Vioxx followed not long after. There were also drugs that modified the autoimmune response: Enbrel and Arava arrived with a lot of fanfare, and a third drug, Remicade, was in the pipeline for approval. Altogether, the new drugs were providing rheumatologists with exciting new tools. Dr. Valen, an investigator for Colloral who practices in LaCrosse, Wisconsin, noted that "this past year has been unlike any other in the twelve years I've been practicing." Of course, the new drugs also meant stiffer competition for Colloral, if it got to market.

On the other hand, Colloral had a lot going for it. It appealed to patients who were averse to side effects and wanted a "natural" solution. Rheumatologists agreed that the real test would come after approval, when they could try it out as an enhancer of other, stronger medications like methotrexate. The challenge was, as Bob Bishop put it one day shortly before the trial ended, to "drag this body across the finish line, one way or another."

Jo Ann Wallace insisted that AutoImmune had really optimized the chances of success with the Colloral trial. "Colloral certainly isn't anything like Myloral," she said. "With Myloral, Bob took a lot of criticism for going into a large phase 3 trial with thirty patients' worth of data. People looked at us and said, 'What are you people smoking?' The drugs themselves are very different. With Myloral you had this Waring blender mishmash of proteins. I mean, I hate to say it, but you were basically putting cow brain in the blender and turning it on high."

Colloral was a more refined drug, and it had been tested in hundreds of patients and found promising. Moreover, the phase 3 trials had been run with almost fanatical attention to detail. "We're doing everything right," Jo Ann Wallace said. "If we don't make it with this study, the drug doesn't deserve to make it."

CHAPTER 16

The End

Thursday, August 26, 1999

Howard Weiner was stretched out in a business-class seat, flying back from a meeting in Brighton, England, drinking red wine, listening to Mary Chapin Carpenter through the earphones, and working on his journal. In less than twenty-four hours, he would know the results of the phase 3 Colloral trial. "I had a dream," he wrote, "that AutoImmune stock was trading at $32—the trial had worked." He imagined what he would do if the results were positive—how he would sing "Happy Days Are Here Again." Tears came to his eyes as he thought of telling everyone in the lab and hugging them.

Then he thought of what it would feel like if the trial failed, and he had to keep going with "this monkey on my back." He remembered riding the roller coaster at Lakeside Park in Denver as a little boy. "As one slowly goes up, one knows soon one will be thrown into the terrifying falls. I am going up the first hill, ready to plummet down."

At midday the next day, Weiner drove out to the offices of Boston BioStatistics in Framingham, where the results of the trial were to be announced. Although he had no appetite, he stopped for a hamburger at a Burger King en route, knowing he should eat something. Even with the stop, he arrived early. Jo Ann Wallace was there, looking calm—a promising sign perhaps—followed by Bob Bishop and finally Joe Boccagno, carrying the acetates that would tell the story. Weiner tried to get a peek and see if there were positive p values, and thought he saw one.

But then Joe Boccagno got up to speak. "Without further ado," he told the small gathering, "let me say that we did not meet our primary outcome measures."

It would take some minutes for that statement to sink in. Everyone in the room tried to deal with the shock, as Boccagno went on calmly to present the data. In essence, he explained, the placebo response "really wiped out this whole study." Colloral did as well as they had expected it to, but the placebo did as well or better. The placebo effect in the Myloral trial had been over 56 percent. The placebo response in the Colloral trial was higher than in any other rheumatoid arthritis trial that Boccagno and his statisticians could find in the literature.

Joe Boccagno, aware of all the hopes, all the work, all the millions of dollars invested in the belief that Colloral would work, knew that words were needed, as they would be in comforting the bereaved. So he talked on about all the ways he and his team looked at the data, to make sure they hadn't missed something. They looked at individual sites to see if any were anomalous. They asked some investigators who were dispensing medication to take their blinders off, to make sure that placebo and drug didn't somehow get mixed up.

They compared this trial to previous trials, and concluded that, in any other trial, the Colloral results would have been statistically significant against placebo. But in this trial, the placebo effect set such a high hurdle that Colloral couldn't beat it. "We had a lot of activity," Joe Boccagno notes. "Patients responded, whether they were on Colloral or placebo."

Bob Bishop, who had been silent through most of the presentation, finally spoke. "That is *mind*-boggling," he said of the placebo effect. "It's testimony to the fact that placebo is real biology."

The placebo effect was the mystery that had haunted AutoImmune's phase 3 trials of both Myloral and Colloral. Everyone—company executives, clinicians, and researchers alike—viewed the placebo as a nuisance, a thing to be overcome, to establish the "genuine" effect of their drug on the body. But in the Colloral trial, the placebo effect was so huge that it demanded attention. A greater than fifty percent improvement on placebo! What could possibly be going on? What was this "real biology" that Bob Bishop always mentioned, in passing, when talking about the placebo effect?

"Placebo," meaning "I shall please," is the first word uttered in the Catholic vespers for the dead. In the Middle Ages, it became associated with professional mourners, hangers-on who were hired to chant at funeral masses. In modern times, it has generally been viewed, similarly, as a necessary evil. Placebo controlled studies began to appear in the 1920s and 1930s. The idea of double-blind, placebo-controlled trials emerged in the 1940s. And in the seventies, the FDA started recommending (and now requires) that studies of new drugs use a double-blind placebo-controlled design wherever possible.

Paradoxically, the increasing use of placebo to buttress evidence-based research has highlighted the mysterious powers of mind and spirit in the healing process. The list of stunning placebo-induced effects is long. The airways of asthma sufferers dilated on a "placebo" bronchodilator. Patients who had a wisdom tooth extracted got as much relief from a fake application of ultrasound as from a real one, providing patient and therapist believed the machine was on. Astonishingly, doctors were able to get rid of warts by painting them with an inert dye and promising patients they would be gone when the color wore off.

Perhaps, as a growing number of researchers have begun to suggest, the placebo effect offers not a roadblock to treatment but an opportunity. "For me," writes Howard Spiro, a gastroenterologist who has turned his attention to placebo, "the placebo provides a lens to look at the stresses in medical practice between science and intuition. Medical knowledge depends upon science, but medical practice requires art as well as science. Study of placebos can span the gap between the two cultures, for its power lies in consolation and suggestion."

One day, Spiro suggests, it may be possible to trace the "billions of uncounted neurons in the human brain" and "patch shattered hopes with enzymatic glue." But for the time being, "mind, thought, and spirit remain more than hard-wired circuits." As David B. Morris, another student of placebo puts it, "Placebos place belief and meaning at the center of the therapeutic encounter."

If placebos are healing, as research powerfully suggests, then the logical next step is to make use of them as a treatment. Some have tried this successfully. At Tulane University, Dr. Eileen Palace has helped nonorgasmic women by hooking them up to a biofeedback machine that measures their vaginal blood flow, an index of arousal. The women are then shown arousing pictures, and told that their blood flow has increased even when it hasn't. Almost immediately the women become aroused.

But it is one thing to play games in the mysterious area of sexual arousal, and another to deceive patients who come into the office with other kinds of health problems. Some suggest that even this can be done; after all, everyone was happy to prescribe aspirin for years before understanding how it worked. Why not give out a little blue sugar pill? But most practitioners view this as unethical. Certainly, such a deception could not be, in the age of "informed consent," the subject of a clinical trial. There are others who argue, however, that it is possible to provide placebo help in pill form without deceiving. Howard Brody, a physician and student of placebo, suggests telling patients of "the mechanisms you have built into your own brain and your own body, that allow you, when you are in the right frame of mind, when you trust, when you have the right therapeutic milieu, to contribute tremendously to the potency of whatever pill I give you. So you, as an autonomous being, have tremendous power within yourself to heal yourself."

But others scoff at such a complicated formulation, claiming most patients don't want all that talk. Despite the mood of the times, when patient autonomy is favored over the old paternalism, there is tremendous power, as Spiro points out, in the words "I will take care of you."

Ironically, the very power of those words, of that promise, may have been one of the things working against AutoImmune in the phase 3 Colloral trial. The attention patients got, in this very well-run trial, was often greater than they had gotten in the entire history of their disease. That alone had a tendency to make them feel better.

Beyond that, other factors tended to inflate the placebo effect. Rheumatoid arthritis, like multiple sclerosis, is a waxing and waning disease. It may be, in both cases, that the patients recruited for the studies came to them at times of more acute distress. As a result, the disease tended to be in a waning period in the months following their entry.

Finally, there is the particular relationship of placebo to pain. Pain is one of the central features of rheumatoid arthritis, and it was an element of three out of the four measures used to assess the efficacy of Colloral. But pain is particularly susceptible to the power of suggestion. Human pain requires the conscious perceiving mind. If you shut it off, through sleep or anesthesia, the pain disappears. As thousands of amputees know, you don't even need a limb to suffer pain in the limb. All you need are the right neurobiological patterns and a conscious mind. The minds of the arthritis patients in the study, which promised a nontoxic oral solution to their pain, had an effect on their bodies even when the drug wasn't present. Colloral may well have helped some RA patients a great deal. But there weren't enough drug responders to surpass the effect of the placebo response.

In the months leading up to the announcement, everyone at the company had been anticipating a positive result. Jo Ann Wallace had writers working on the FDA application, and lawyers looking into over-the-counter competition that might infringe on Colloral's patents. Enough Colloral had been stockpiled to keep the patients in the trial supplied until the FDA approved the drug. Everyone had been operating on the assumption that the trial would succeed.

Now, all that forward motion had been arrested. It was the end, as everyone in the room knew, of AutoImmune as a company with employees and offices. There would still be patents, and the hope of profiting from some future breakthrough, made somewhere else. But the dream of AutoImmune as a successful company ended when Joe Boccagno uttered his first sentence. Now the challenge was to shift direction, and start an orderly dismantling of the enterprise.

Bishop reminded the group that they must keep the news to themselves until the following Wednesday, when the announcement of the results would be made public. "We did everything we needed to do," he told them. "I don't have any of those feelings in the back of my mind that 'if only we had. . . .' And it is what it is."

Howard Weiner rose to tell the others that he recently attended a mucosal immunology meeting in Amsterdam where Chuck Elson, a sometime critic, predicted that oral tolerance would be successfully applied to human disease in the twenty-first century. "The question is," he added, "how is it going to happen?" And then regret began to seep in. "I don't know, it's hard to understand—How many more phase 2 trials can you do before you do phase 3?"

"It is like a death in the family," he told them. "But it's also different, because everything we do here lives on, even when we're no longer here."

After the meeting, Weiner went on to the lab, feeling numb, and told people there the news. Later in the day, he went to the airport to pick up Mira, who had been in Europe on business. He had agreed to give his wife a thumbs-up or thumbs-down the minute they found each other. But no signal was necessary when she saw his face.

"It will pass," he wrote in his diary later that day. "A certain amount of time must pass."

In some ways, the failure of the Colloral trial was even harder on Bob Bishop than on Howard Weiner. Weiner's commitment to oral tolerance and to finding a cure for MS would live on after the company faded. Bishop's heart and soul had been poured into the success of the company. Seven years earlier, he had uprooted his family and given up his safe job to gamble on AutoImmune. For Susie Bishop, it had been the most difficult move of her married life. Until then, she had been an enthusiastic executive wife, and couldn't understand why other women in similar positions seemed bitter and unhappy. After a hot, lonely summer, far from California, she understood much better. She had made Boston work, but it had been a struggle.

Bob Bishop had been optimistic and enthusiastic from the moment he took the job. He had never stopped believing. Whenever you passed his office, he would say something optimistic, or raise both hands with fingers crossed. "He's our cheerleader," Jo Ann would say. He was the one who worked to keep everyone's morale up through difficult times.

As the day of reckoning approached, Bishop thought about AutoImmune all the time, except when he played golf. He dreamed about it many nights. But now, with the news of the Colloral trial failure, all the air suddenly went out of his balloon. He was so despondent, so unlike himself, that Jo Ann worried about his driving home alone from Framingham after the announcement. Howard asked him if he wanted to go out for a drink, but he said he just wanted to go home to Susie.

Bishop and the AutoImmune executives learned the results of the Colloral trial on a Friday. On the following Tuesday, Joe Boccagno brought the same news to the AutoImmune board. He ended his report, which took only fifteen minutes, by asking if there were questions. No one spoke at first. Then Henri Termeer, CEO of Genzyme and board member, said it was pretty hard to have any questions after seeing that data. It was so clear-cut. The board members worked together on writing a press release, and discussed what to do with the $6 million AutoImmune had left. Should they try to market Colloral as an over-the-counter drug? Should they return the money to the stockholders? Or should they try to join up with another venture? All of that was left to the future. For the moment, the one thing that was clear was that AutoImmune, which had started out as a virtual company, was about to become virtual again. And the person who would have the unenviable task of making that happen was Bob Bishop.

The press release, which went out the next morning, emphasized the unusually large placebo effect: "Substantial improvements from baseline were noted in the Colloral treated group," it stated. "Unfortunately, the size of the placebo response was much greater than previously observed."

Then there was a quote from Bob Bishop: "We will immediately reduce our headcount and other operating expenses to conserve resources as we evaluate our strategic options," followed by a list of all the good news: an extensive worldwide patent portfolio, two externally funded clinical trials in diabetes, one in organ transplant, and an agreement with Teva in connection with the oral Copaxone trial in MS. "AutoImmune will receive milestone payments and royalties once products are approved," the release predicted optimistically, and went on to quote Howard Weiner, professor at Harvard Medical School, expressing confidence that "with increased knowledge, the application of mucosal tolerance for the treatment of human diseases will be realized."

As soon as the release went out, share values dropped to below a dollar. Newspapers the next morning picked up on the Colloral failure and little else in the press release. "Troubled AutoImmune Shares Take a Beating," read the headline in the *Boston Herald*. The *Wall Street Journal* ran a brief story under the heading "Stock Plunges 74% as Tests of Drug Show Poor Results." The *Boston Globe* headline delivered a one-two punch, citing the preliminary success of another company's arthritis drug: "AutoImmune falters," the headline read, "Vertex soars."

The *Globe* quoted Bob Bishop sounding upbeat in the face of adversity: "I don't have any second thoughts or nagging feelings over what we did or how we did it," he told the *Globe*. "This is just one of the vagaries of pharmaceutical development, but we have to move forward. I am proud of the people and team that we put together."

Speaking to a reporter he barely knew, Bishop managed to hang onto his supersalesman persona. But it wasn't so easy to do at the meeting that Wednesday morning with the twenty people who were still on the AutoImmune payroll.

Even before the meeting, the AutoImmune employees had gotten a copy of the press release in their boxes and knew about the disappointing results. Everyone lingered outside the small meeting room, joking halfheartedly and munching on the doughnuts and coffee that had been provided. They were reluctant to hear the details.

Slowly, the chairs filled up. Joe Boccagno was seated in the front row, ready to do his job, as was Howard Weiner. And Bob Bishop was standing up front, wearing his executive uniform: starched white shirt, yellow-and-black tie, gray flannels, and shiny black loafers. His large frame was half perched on the edge of a rectangular table. "I didn't want to be doing this with you today," Bishop began, and went on to discuss the results.

"The placebo effect was two and a half times greater than before," he noted. "We would have had success in three of the four endpoints required for FDA approval, if the placebo rate had stayed at the previous level."

After going over the results and answering a few half-hearted questions, Bishop outlined what needed to be done. "We need to get small as quickly as we can." That meant canceling all contracts, calling vendors and getting final invoices, tossing out chemicals and unnecessary files. "Final checks, including severance," he told them, "will be available tomorrow for most of the people in this room."

Then Bishop shifted to a more personal level. "I'm proud of you. I'm proud of you as individuals and as a team. And I know you're going to do fine whatever you do."

There was a long silence, as Bishop tried to suppress a sob that came welling up.

He made a feeble joke: "I don't know why I didn't have this trouble with the board."

He put his hand over his mouth, and the barely audible sobs made him unable to speak for what seemed like a very long time. Finally, he gave up trying. "I think you all know how I feel," he said weakly, and sat down.

Someone in the audience murmured, "I think we do." A number of people in the audience were crying. Even Jo Ann Wallace had tears running down the front of her black knit shirt.

Howard Weiner got up and gave a brief history of the oral tolerance endeavor: the discoveries in the early eighties, the company with no name, the first meeting with Bob where they talked over lunch—here he gave Bishop a big smile—and the growth of the field of research. Insulin failed in its first trial, he reminded them, and so did penicillin. "Other people," Weiner told them, "are going to make this work based on what we did." Even though he was shocked and saddened, he was at peace. "I don't think anything could have been done differently," he concluded.

By the end of Weiner's speech, Bob Bishop was on his feet again. "I'm recomposed," he told the audience, and proceeded to talk about what could be "salvaged" of the company.

Minutes after the meeting ended, a big canvas trash bin on wheels appeared in the hallway outside the offices. Within half an hour, it was full to the brim with loose paper and shiny white three-ring notebooks.

Despite the statistics, there were individuals who were helped by Colloral. For them, the news of the trial failure was deeply distressing. Jo Ann Wallace had many calls protesting the discontinuance of the drug, including one from a man who had suffered with rheumatoid arthritis for twenty years, and who was unable to dress himself. After he took Colloral, his arthritis was nearly gone. A patient of Lea Sewell's who had been in an early trial had been taking Colloral since 1993. When he started it he was "about ready to go into a wheelchair." Now, though he had flares from time to time, he did karate and water aerobics. And then there was Manual Barboza. Arlene Turgeon called him and told him, "I've got bad news." She told him to write a letter to the company. "So I wrote them a letter and told them how good it was doing. And to let me know if I could get it, and to let me know where."

And if he can't get it?

"What are you going to do?" he asks with a shrug. "You kind of go with the wind."

Bob Bishop and Howard Weiner saw quite a lot of each other during those trying days after the announcement: They played golf together, badly, and they went off on the weekend with their wives to the Mt. Washington hotel, a huge old resort hotel in the White Mountains of New Hampshire.

The following week was Rosh Hashanah, which afforded a chance for Weiner to be with his sons, and even to shoot baskets with them on his restored leg. Then that Sunday, Bob Bishop joined the poker game at Howard's house. Bishop had played in Howard's poker game before, so he knew about their tradition of dealing an extra hand, in one of the games, to an invisible player they called Elijah—after the unseen guest at the seder meal. Elijah's cards lay face-down on the table until one of the group won a round. Then, the winner pitted his hand against Elijah's. If the flesh-and-blood player's hand trumped Elijah's, he got to haul in the chips that were piling up in the middle of the table. If Elijah won, then everyone just anted up again and the mountain of chips in the middle of the table grew larger.

As it turned out, Elijah was in rare form on the night, shortly after the Colloral results came in, that Bob Bishop joined the poker game. Many times, the little gathering of doctors and writers, along with the CEO, challenged the unseen hand and lost. But finally, at the end of the game, Bob Bishop beat all the players and Elijah and hauled in all the chips. The group never played for big stakes, so Bishop's total win came to no more than $70. Still, on most nights, it would have been reason to crow. But when the others congratulated him, Bishop told them what they already knew: He would rather have had his good luck in the arthritis trial.

CHAPTER 17

The Beginning

"SUCCESS IN SCIENCE," according to one of Howard Weiner's aphorisms, "is going from one failure to another without losing enthusiasm." Even though the failure of AutoImmune Incorporated looked pretty final on the pages of the *Wall Street Journal,* the ideas on which the company was built continued to have resonance. Another aphorism that could have been on Weiner's wall was even more appropriate to the moment: the only failed experiment is the one you don't learn from. AutoImmune's failed trials had many lessons to teach about oral tolerance as an approach to autoimmune disease and about the process of moving from scientific discovery to clinical application.

Henri Termeer, CEO of Genzyme and member of the board of AutoImmune, predicted that, sooner or later, someone would get oral tolerance to work in humans. Success, when it came, would build on the results of AutoImmune's failed trials. "This work will continue to fascinate people, because it's such a very nice way to treat. Tremendous progress was made, even though the company did not survive."

Termeer's Genzyme is one of the companies that always makes the shortlist of biotech successes. Yet no one is more aware than he that every success is built on many failures. He gives gene therapy as an example. "There have been many gene therapy companies that have failed. But I don't think there are many people in the industry who believe that gene therapy will not ultimately become successful."

His own company has worked for ten years on applying gene therapy to cystic fibrosis. "We did about seven trials for cystic fibrosis and weren't able to get to something we could take to a formalized product development program. But we learned, we learned a lot about vectors [a vehicle used to insert a piece of DNA into a cell]. As a result we've changed the program direction from chronic lung disease, where these vectors are a disadvantage, to acute disease where they can be useful. So now we are in clinical trials in cancer and in cardiovascular disease. You could say we failed on the cystic fibrosis program. It wasn't successful, but it was not a failure. We learned from it."

By this measure, AutoImmune wasn't a failure either: Nearly everyone involved in the enterprise came away wiser, in one way or another. For Henri Termeer, the failure affirmed his belief that biotech research needs to be done differently in the future.

"Biotechnology has been good for innovation," Termeer observes, "but bad for investment. Every year for twenty years, biotechnology companies have lost more money than they did the year before." Termeer's own large office, furnished in leather as soft as a baby's cheek, is located on a high floor of a nondescript cinderblock building not far from MIT. But Genzyme's signature building, a whimsical brick structure with shiny peaked roofs and gables along the Charles River in Boston, is a pleasing oasis in the midst of one of Boston's ugliest intersections of highway and turnpike.

Termeer himself is as elegant as his building, a small man in impeccably tailored suits. He is in his fifties, with dark red hair that appears to be his own, a rosy complexion, and eyes buried deep beneath folds of skin. A native of Holland, he got his start at Baxter, one of the giant pharmaceuticals. But he speaks with the most excitement of the early days of biotech, when he would spend his weekends wandering around MIT with scientists, dreaming up new possibilities.

Genzyme, under Termeer's leadership, has been particularly good at identifying small patient populations on which the big pharmaceutical companies are unlikely to concentrate. Their best-known success is a drug to treat a rare genetic illness called Gaucher's disease, a painful hereditary disorder that results in an enlarged spleen and liver, as well as bone abnormalities. Two more drugs, related to kidney transplant and thyroid cancer, recently went on the market. Genzyme casts a wide net in its research, focusing not just on drugs but also on diagnostic tools. "Our objective was to build a company that was sustainable," Termeer explains. "That was a choice, and there are pluses and minuses. It's a lot of work and a lot of people."

AutoImmune Inc. was the opposite of the multifaceted Genzyme. It was what is called a "pure play" company, based on a single promising idea. "It was an oral treatment that would be nontoxic and easy for patients, in a disease that's very serious. The manufacturing of the product was very easy. And so you could get into the clinic relatively inexpensively. And they just placed that bet." Termeer, as an AutoImmune board member, thought it was a good bet at the time. "Hindsight would say this was ill advised. But a pure play, without wasting your time, is an attractive proposition."

Termeer thinks now that the mistake AutoImmune made was probably in the timing. Some criticized the company for rushing too quickly into phase 3 trials of their MS drug, Myloral. But Termeer believes that the mistake occurred earlier, when the decision was made to transfer the oral tolerance research out of academia. "It was the first decision to set up this commercial enterprise. That should have been done much later."

The problem, as AutoImmune executives would be quick to point out, is that clinical trials are enormously expensive, and sufficient

NIH funds for trials have been hard to come by. Termeer would agree, but points out that that is changing: Owing in large part to lobbying by the research community, the NIH budget is on track to double in the next five years. "With NIH funding," he says, "these technologies can stay much longer in academia." Ten to fifteen years, the time it often takes to bring a promising drug through the trial process to positive results, is too long, in his opinion, for a commercial company to tolerate unprofitability. Private capital simply won't stick around. Indeed, one of the reasons for the sag in biotech is that "venture capital has wised up and has other places to go with their money—like the Internet." Since Termeer made that statement, of course, the Internet has turned out to be dicey as well.

The failure of AutoImmune, in Termeer's opinion, was part of a realignment in biotech. "There are going to be some of these pure play companies that are very exciting. And some of them may hit a bull's-eye. But the number of pure play companies will decrease—they are decreasing as we speak." He notes that Alpha Beta and Immunologic, in the Boston area, took a dive at around the same time AutoImmune did. He predicts that research will stay in academia longer in the future, and companies that do emerge will be larger and better capitalized. "The revolution in biotechnology will continue, except it will be technology in an academic sense."

For Barry Weinberg, too, the failure of AutoImmune contributed to a reassessment of biotech in general. The limited partners in the CW Group actually made money on the undertaking, since they were able to buy the stock at $3 a share and sell it, after the public offering, at around $16. Weinberg lost money in his own stock portfolio, though he dismisses his loss with a shrug. "That's the risk," he says. Nevertheless, the AutoImmune failure contributed to a new realism about biotechnology.

"In order to be in the venture capital business," Barry Weinberg points out, "you have to be optimistic. But I think that we were overly optimistic. We all thought that the advent of some of these biotechnology tools was going to dramatically reduce the amount of time it took to get a drug through the development process and into the clinic. But really these tools have not reduced either the risk, the time, or the cost of bringing the product to market."

One of the companies Weinberg is working on now, called Fast-track, uses computer analysis to try to predict more accurately the outcome of clinical trials. Numerous other current investments have to do with information technology in health care. As for traditional biotech companies, the CW Group has invested in only one in the last six years. "I think we are one of the few health care firms still willing to make investments in biotech. But the candidates just don't meet our investment criteria."

In the 1970s and 1980s, when "biotechnology" was a new word and a new concept, venture capital poured in to what seemed like an almost sure thing. The traditional way to develop drugs, using chemistry to copy nature or to discover active molecular combinations, was a game of chance. Biotechnology had the potential to replicate the stuff of life. When nature erred, biological science could make adjustments, using recombinant technology, or so early investors hoped.

But research requires great patience, and capital is restless. Not surprisingly, the marriage of the two in biotech companies has been rocky. There were spectacular biotech successes in the early days, and their names are often cited in the industry: Amgen, Biogen, Genentech, Genetics Institute, and of course Genzyme. Altogether, about ten companies made it early on, most by coming up with proteins to replace defective ones, and they created enormous excitement and generated a lot of investment. By the early nineties, biotech companies were appearing at the rate of two or three a week.

But long before the Colloral trial failed, the air had gone out of the biotech balloon. When Jo Ann Wallace ran into an investment banker at a party early in 1999 and told him she worked in the biotech industry, a look of pity and alarm came into his eyes. "None of us want to talk to biotech," he told her.

As long as biotech was hot, investors could make money in companies that were a long way from making a profit, just as they did more recently with the Internet. But the complexity of drug development has, over time, dampened Wall Street's enthusiasm. It's not just a question of producing and selling a product. There are regulatory hurdles to jump, and they require the cooperation of the sometimes sluggish FDA. Patent protection is complicated as well. Finally, and most important, there is Mother Nature. As Howard Weiner often reminds his people, "you have to go by the biology." And more often than not, as he also is likely to say, "it takes time for the biology to declare itself."

What's more, as another early investor, Alan Ferguson of 3i noted, "biology isn't any better yet at solving the problems than chemistry was, when the big pharmaceutical companies were the only ones working on it."

—o—

The employees of AutoImmune responded to the breakup of the company with varying degrees of disappointment and resignation. For Orysia Komarynskyj, who had thrown herself into the direction of the Colloral trial, the disappointment was enormous. "We believed in this oral tolerance," she explains. "It was a belief that I would say was almost religious. And I've never experienced that before—it was for those patients out there." Others were more matter of fact: "This is a risky business," regulatory affairs director J. D. Bernardi points out. "This is not selling Clorox bleach. It's also a fascinating business. I enjoy it. So I'm not averse to the risk."

"There will always be people out there," Bernardi notes, "with new ideas and new projects and new therapies to help mankind, so I'm not worried that I won't have a job. I may be out of a job for six months, but I may also hitch onto one of the stars and ride it up."

Bernardi was right about the jobs. Whatever their reaction to the demise of the company, AutoImmune employees didn't have to wait long for new opportunities. Bernardi signed on at a company called Praecis. Joe Boccagno quickly found work at Transkeriotics, a Cambridge biotech, as a director of clinical research.

By the time of the Colloral announcement, Malcolm Fletcher's life had done a flip-flop. Instead of living alone in a small apartment and making himself shepherd's pies, he was installed with his wife in his new house, the Churchill model, and there was hope that the custody dispute over his wife's children, the reason he had left AutoImmune early, would finally be resolved. His personal life, he reported, was much improved, though his work life was the opposite. "I have a job that looks fine on paper," he reported, "but doesn't come close to AutoImmune."

Jo Ann Wallace had always been matter-of-fact about the possibility of failure, and—except for a rough patch on the day after the Colloral announcement—she remained that way. Even a $40,000 loss with the decline of the AutoImmune stock didn't bother her too much. But she also wasn't eager to start in right away at a new job. For one thing, her partner, Bobby Owen, was still employed by a struggling biotech in the Boston area, and she was determined to stay close by. "At my age," she said, "I'm not interested in a commuter marriage for six months. I've done it four times in twenty-five years and I'm not gonna do it again."

Even though she claimed she was "not a good playperson," there were a lot of things she wanted to do. Her leg, fractured badly at one of the lowest points in AutoImmune's journey, was completely healed, and she returned to golf with the intensity she used to devote to the company. One week in the spring of 2000 she played thirty-six holes four days in a row. She also spent a lot of time pursuing good food and wine, traveling several times to Chicago on wine-buying trips.

Bob Bishop, in the weeks after the Colloral announcement, seemed to be more energetic than ever as he worked to wind up business at the company. Only someone who had known him a while would notice that he was talking a little too fast, saying all the right upbeat things, but sounding like he was on automatic.

There was a cartoon, drawn by a friend, on the wall of his soon-to-be-vacant office that probably expressed his true feelings better than he could. It was a caricature of him sitting at his desk, smiling crookedly. Hanging on the wall behind him, in the cartoon, was a gun under glass, with the instructions "Break Glass if Clinical Trials Fail."

Over time, Bishop grew calmer. "Our attitudes are fully adjusted," he reported ten months later. He and his wife Susie returned home to California and bought a 1930s vintage house high in the hills of Pasadena, overlooking the Rose Bowl. He continued to spend about twelve hours a week on AutoImmune business, monitoring the progress in Lilly's ongoing diabetes trial, as well as the very long but promising trial that the NIH was sponsoring in juvenile diabetes.

Among major stockholders, Bishop said, no one was asking for "liquidation." Rather than collecting the $0.40 a share their stock was worth, they preferred to "preserve the upside," believing there was still a possibility that one of AutoImmune's remaining ventures would succeed. In this connection, there was more good news than bad.

AutoImmune's patents on the discoveries coming out of the Weiner lab provided some support for current work and considerable hope for the future. After the failure of the Colloral trial, the AutoImmune team worried that Teva might give up on their trial of Coral, the oral form of Copaxone. Bob Bishop traveled to Israel to meet with Teva executives to spur them on. But it turned out that the Teva team had no intention of giving up, and was going full speed ahead with, and even expanding, the Coral trial. Teva's trial would take place at 158 sites in eighteen countries around the world, with final results coming out early in 2002. Because of patent claims, a successful trial could yield $50 million a year in royalty income for AutoImmune. If that happened, AutoImmune could rebuild or—the more likely scenario—be purchased by another drug company for a respectable price that would reward shareholders. "I'm hopeful," Bishop said. "It's the strongest animal data we've had."

In addition, AutoImmune struck a deal with Elan Pharmaceuticals in the spring of 2000 in connection with the development of a drug for Alzheimer's disease. Alzheimer's, though not an autoimmune disease, has an inflammatory component: it is caused by toxic shards of protein called amyloids that build up in the brain. In 1994, the Weiner lab, working jointly for the first time with Selkoe's Alzheimer's group across the hall, began to explore the possibility that amyloid buildup could be decreased by oral or nasal administration of the substance. At the same time, Elan was trying another approach, injecting amyloid. Elan published positive findings first in *Science* in July 1999. But the Weiner-Selkoe lab's experiments were also yielding promising results. And in the spring of 2000 Elan agreed to pay AutoImmune a total of $7 million for its patent on the mucosal approach. "It will allow Elan to make an improvement downstream," Bob Bishop explained. "Also, and most important, their having this patent will prevent someone else from chasing them."

For AutoImmune, the $7 million, in three payouts, from Elan, added to the $6 million still on hand, meant that they could become, in the language of business, "cash neutral." There was enough, from interest on the principal, to pay fees on patents, with a little left over for Weiner's continuing research.

"We're not quitting," Bishop said, "at least not yet. When we started down this road, 10 percent of the immunologists knew what oral tolerance was. And none of them thought we had any chance of being successful. Now 90 percent of people in the field know what it is, and most of them say we just haven't figured out quite how to make it work yet. But it will happen."

Howard Weiner's reaction to the Colloral trial failure, after he got over his initial disappointment, was to ask the obvious next question: what went wrong? In his journal, he ruminated on the failure of both the Myloral and the Colloral trials. First of all, in the Myloral trial, it was probably a mistake to give patients a mixture of myelin proteins from cow brain. It would have been better, as some of Caroline Whitacre's subsequent research confirmed, to give them one protein, probably myelin basic protein (MBP) and rely on the bystander effect to activate suppression. But at the time he and the AutoImmune group set up the trial, bystander suppression wasn't fully understood: As so often happens, the research advanced, but the protocol for the clinical trial couldn't be changed to take advantage of the new breakthroughs. The protocol was frozen in time.

In fact, it probably would have been best not to use cow brain at all, but rather to produce human MBP through a recombinant process. But when Weiner and Barry Weinberg considered that possibility, early on, they decided it would be too slow and costly. Once again, the ticking of the investors' clock affected decision making.

Weiner believed, looking back, that there should have been a dosing trial of Myloral before the phase 3 trial. But even if there had been, it would have been difficult and time-consuming to establish the efficacy of different doses, because there was no "surrogate marker" that could signal the activity of the disease or show the effect of the drug on the immune system. Since the Myloral trial, it has become easier to trace the progress of disease in MS patients using MRI in combination with gadolinium, a radiologic marker that lights up new lesions. But even the MRI with gadolinium is imperfect. The failure of the Myloral and Colloral trials made Weiner even more eager to find a true "surrogate marker" in the blood, which he views as "the key to immune manipulation in the future." If it were possible to take a sample of blood and measure disease activity in it, as we do with infection, then the process of testing the efficacy of drugs for autoimmune diseases would be greatly simplified. At the CND, work on finding a surrogate marker, both for autoimmune disease and oral tolerance, has intensified since the failure of the Colloral trial.

The results might have been different too, Weiner now thought, had the trial substances been introduced through the nose rather than the mouth. Current research suggested a stronger impact with nasal preparations in some instances. Research in animals also suggested that it might be necessary to add an adjuvant, some potent substance that would boost the effect of oral or nasal tolerance. All of this pointed the way, for Weiner and others at the center, toward what needed to be explored next.

For the researchers working on oral tolerance in the Weiner lab, the failure of AutoImmune was disappointing, but it didn't alter their general direction. The work of the principal investigators preceded the beginning of AutoImmune, Inc. and would extend beyond its demise. The postdocs, in most cases, would go on to other labs and work on other questions, connected in one way or another to the oral tolerance research they had done in the Weiner lab. Usually their career paths were influenced by other pressing issues in their lives as well.

Gabriela Garcia, the Brazilian who had worked on mouse arthritis and oral tolerance in the Weiner lab, decided that she wanted to stay in the United States. To get permanent resident status, she needed to find a job in the private sector. She had hoped to work at AutoImmune, but when the company faltered, she had to look elsewhere.

Out of several possibilities, she chose a job with Astra, a Swedish company with a laboratory in Cambridge that has since merged to become AstraZeneca and moved to Waltham. She liked the benefits they offered, which compared favorably with those offered by U.S. companies. "The Americans want to squeeze the workers as much as they can," she explained. Even more important, she was interested in the work—a search for a vaccine against *heliobacter pylori* bacteria *(h. pylori),* which cause ulcers. It wasn't oral tolerance, but it involved the mucosal system, specifically the stomach lining, which she had worked on from her earliest days as a student of the great immunologist Nelson Vaz in Brazil.

Gabriela found her job in biotech to be a pleasant surprise. In academia, she explained, everyone was focused on publishing. To be first author on a paper, you needed to do the work yourself. That meant, in her case, that she was always overwhelmed by the workload. "You were responsible for *everything!*" she remembers. "And I think that's not very productive. The pressure is unbearable. There is only one day that you kill the mice and you have to take all the tissues that you want and do all the experiments. And all the information that you should take out from each experiment you cannot, because you are alone. So you keep repeating experiments."

At AstraZeneca, Gabriela is part of a team that includes an American boss, a colleague from Russia, and another from Hungary. All of them have their own technicians, which greatly relieves the pressure. In addition, on the day that they sacrifice mice, everyone on the team works together. "I don't want to say that all companies are like this," Gabriela adds. "It depends on the lab."

One reason for the difference, she believes, is that the goal of the company is not publication, even though company scientists want and need this. The goal is to make a product, in this case a vaccine against *h. pylori*. The competition is fierce. Another company, just down Sydney Street in this biotech-rich neighborhood of Cambridge, is working to develop a vaccine too. "There is a pressure," she acknowledges.

Also, there is less freedom to share information. When Gabriela gives a paper at a conference, there are certain things the company won't let her say. Recently, when she presented a poster at an immunology conference in India, university scientists wanted to know what agent she was using in her experiment. "And I had to tell them I cannot say. It's not very comfortable."

Nevertheless, Gabriela doesn't feel that the pressure to get to the product affects her day-to-day work at the bench. "The framework is that we want a vaccine. But inside the frame, it's my group that decides what experiments to do." She adds: "I'm working in a company that is very rich, so they can afford to look into the processes which are important to know, but not necessarily important to get to a product. I don't know how it works in small companies."

Unlike Gabriela Garcia, Anthony Slavin decided to stay in academic research. But he moved on from the Weiner lab, in the summer of 1999, to the Stanford lab of another prominent researcher, Gary Fathman, where he would work on autoimmune mechanisms. For him the Weiner lab was "sort of like your first girlfriend. You never quite forget. And Howard's been very good to me over the years." Anthony had a real girlfriend by the time he left—a woman he met in the lab when she was working there as a technician. He decided to go to California because she was out there, getting a master's degree so that she could teach science.

Ruthie Maron stayed behind. Ruthie continued to test new possibilities for the use of oral tolerance, with the help of a new team at the center. Because of her work, and work in other labs, the number of diseases that responded in animal trials continued to grow. Ruthie had no plans to move on. "I'm stuck," she said. "But happily stuck."

She added: "I wouldn't stay unhappily stuck for long."

The truth is that Ruthie Maron, far from being stuck, was at the center of several exciting developments. In October 2000, positive results from the joint work of the Weiner and Selkoe labs on mucosal tolerance as a treatment for Alzheimer's disease were published in *Annals of Neurology.* In the mouse model, a synthetic form of the amyloid plaques that cause Alzheimer's was introduced nasally and had the effect of slowing plaque formation. Plans are now under way for a phase 1 safety trial of this "nasal vaccine" approach in humans.

The Weiner lab was also involved in animal experiments involving oral tolerance as a treatment for atherosclerosis in which the preliminary results were positive. Also, the diabetes trials at Lilly and NIH were ongoing. NIH researchers also reported positive results in animal trials in stroke.

At the same time, other labs were getting positive results using oral tolerance. Researchers at the University of Tennessee found that oral administration of bovine collagen had a significant effect on some of the signs and symptoms of systemic sclerosis, a disease that involves degenerative changes and scarring in the skin, joints, and internal organs.

In November, just six weeks after the failure of the Colloral trial, Norman Staines, who had pioneered the use of oral tolerance in animal studies of rheumatoid arthritis, was one of a group from London who reported statistically significant results in a trial of humans with rheumatoid arthritis that bore a strong resemblance to the Colloral trial. In this small study, fifty-five patients were evaluated. There was a placebo group and three treatment groups, each receiving a different oral dose of collagen. The group who received the middle-size dose of collagen improved modestly but significantly. Howard Weiner had said, on the day the Colloral trial failed, that "other people are going to make this work." It was already beginning to happen.

For Weiner, in addition to the continuing work on oral tolerance, there were two exciting new undertakings after AutoImmune collapsed that kept his mind from dwelling on the disappointment. One was the new MS Center. The other was the trial of oral Copaxone, or Coral. The Coral trial took advantage of what was learned from the failure of the Myloral trial. Unlike the Myloral trial, which used only one dose of drug, Coral would be compared against placebo in two different doses, one of 5 milligrams, and one of 50 milligrams. Second, Coral was a synthetic substance made of well-defined proteins. Animal research suggested that it was more effective than Myloral, a complex mixture from nature containing active and inactive ingredients. Also, the Coral trial would use a more elaborate method of blinding the physician, with the hope of coming up with clearer results. There would be a treating neurologist, to take care of the patient, and an examining neurologist, who would conduct the neurological exam. Finally, the Coral trial would take advantage of advances in MRI technology, such as the use of gadolinium, that hadn't been available when the Myloral trial began. Because of the MRI advances and the numbers in the study, the Coral trial would be half as long as the Myloral trial, lasting one year instead of two.

A successful Coral trial would be a vindication, Weiner believed, of his belief in oral tolerance as an approach to human autoimmune disease. Also, quite apart from the scientific implications, Coral's success would be a major business coup for Teva. As everyone in the pharmaceutical industry knows, a pill will win out over injections every time. As Pravin Chaturvedi, director of pharmacology for Vertex Pharmaceuticals, notes, "We absolutely abhor injections, we hate suppositories, we even dislike big pills. What we like is a small white tablet in a blisterpack that we can take in the bathroom without anyone noticing."

"If Coral works," Jo Ann Wallace quipped, "that's when you short Biogen."

If Coral becomes available, Myloral trial patients will be among the first to line up. Since the Myloral trial ended, Karyn Kraft has debated about going on one of the ABC drugs and has decided against it. Her doctor suggested Avonex, but she was reluctant to try it because she knew someone who took it every Friday so that he could have the side effects on the weekend and be ready for work Monday morning. "What's the point of that?" she asks. "Also it would have cost me $200 a month." Copaxone, which is supposed to have fewer side effects, was a possibility until she saw the warnings on the video. "That was worse than the Avonex," she claims. She also disliked the idea of injecting herself. Recently, she had started taking an antioxidant made from pine bark, highly recommended by some others with MS. But what she'd really like is to get Myloral back. "I'm still sold on that stuff," she says. She would take Coral if she could.

Karyn keeps her cane by her side at all times now, and she has had two bad falls in her kitchen. Recently, when she went to Patriots' Day ceremonies on a trip to Boston, she got so cold and wet that she was unable to move, and had to ask the ranger to drive her back to her car. But she is resilient. "I haven't broken any bones," she points out. She is still working full-time and shepherding her daughter Lindsay through the transition to high school. She has acquired a dog, which takes some of her attention away from Lindsay. "That's good because Lindsay is now fourteen and she doesn't need to be watched all the time."

After Myloral, Nichols H., the Greek-born graduate student living in Montreal, went on Betaseron, the only one of the ABC drugs available in Canada, and was doing pretty well for a while. "I had absolutely no problem injecting myself," he says, "but about a year ago, all of a sudden, I just couldn't do it any more." Then he switched to another interferon, not available in the United States, that came with an autoinjector. "This facilitates the whole process," he explains, "but I still would very much prefer an oral preparation." The good news, Nicholas adds, is that he has been attack-free since he started the interferon regimen.

Allison Hay, the insurance worker in Allentown, Pennsylvania, might be one of the few who doesn't mind giving herself an injection. She is on Copaxone, which requires a daily injection under the skin. "I inject every night," she says. "It's the tiniest little needle and syringe. It's just part of your routine, like brushing your teeth." She likes the idea that the medicine "goes into my body every single day." Even if Coral works, she says, she's not sure she would be in a hurry to change to it.

Allison has good reason to be so enthusiastic. In the year plus a few months after the Myloral trial ended, she acquired three new lesions, making a total of five. Then she started on Copaxone.

Since then, she's been feeling almost entirely well, except for two episodes of weakness and numbness. And she has now become an advocate for Teva, traveling around the country and "encouraging people with MS to initiate injectible drug therapy." She's still planning to have a baby, she says, adding, "I'm thirty-five and the clock's ticking."

For about ten years, Howard Weiner had been entertaining the idea of a new MS Center that would integrate all the elements required for studying and treating patients with multiple sclerosis. "We had our clinic at the Brigham and Women's Hospital" he explains, "but didn't have our own space. We needed to take it to the next level." At the new center, everything would be in one place: examining rooms, infusion rooms, physical therapy rooms, the labs. There would be an MRI machine dedicated to MS research, allowing Weiner and his group to gather images of the brains of 1,000 MS patients over time, to study the course of their disease.

Weiner raised over $4 million from private sources for a center. But money wasn't enough to secure a space in the full-to-bursting Longwood medical area. For two years, Weiner met with the space people about his MS Center and got nowhere. Then one day they showed him a large area above the medical school bookstore, in a stylish pink stone and red brick building across from the hospitals. Weiner immediately knew it was right. But the administrator who was showing him around said he wasn't sure they could give it to him. That was when Weiner surprised himself.

"I kind of threw a tantrum," he says, smiling a little sheepishly. "Not a real one. But sometimes you've got to get mad. I yelled and I screamed and said you have to give it to us." He cited the longevity of his MS clinic, he pointed out that he'd raised the money, he suggested that some of his donors were on the verge of taking their money back. He even brought up a dying member of the hospital board who had been a big supporter and wanted to live to see the MS Center become a reality. "The tirade lasted," Weiner remembers, "for forty-five minutes!"

Samia Khoury, who had come along to look at the space, was astonished. She had never seen Howard Weiner explode in such a way. Most of the time, when he's upset, Weiner goes silent. Sometimes he'll literally absent himself. Or he'll complain in private. Once before, when he was in a battle for lab space, he spent an entire weekend walking around the house swearing to himself. So for Howard Weiner, the tantrum was a breakthrough of sorts. And it worked. That, on top of the business plan and the $4 million, got him the space.

The new MS Center opened to patients on April 25, 2000. At the staff meeting that day, nurses and doctors paused briefly, in the midst of a discussion of all the glitches, to raise plastic glasses filled with champagne to the accomplishment. "This is where we'll stay," Weiner declared, "until we find a cure for MS."

By Monday, June 18, the MS Center was operating fairly smoothly. There were fresh flowers on the reception desk, and a team of helpful receptionists behind it. The dedicated MRI machine was set up on the basement level of a nearby building and would soon be sending its images across the way to screens at the center. The examining rooms and offices were furnished, the infusion rooms were in use. There was even a strip of maroon rug, exactly twenty-five feet long, inlaid in the gray carpet of one hallway, so that the neurologists on the team could use the ambulation index to assess MS patients' walking ability. There was also a video camera in place, so that the progression of individual patient's diseases could be documented for the long-term study the center was undertaking.

That day, Howard Weiner had only one patient to see. She was a forty-seven-year-old woman who had come in a month earlier because she was having trouble with her eyes. She could see things on the periphery of her visual field, but nothing in the center. With the help of MRI enhanced by gadolinium, she was diagnosed with MS. Soon after, she was approached about joining the trial Teva had undertaken to test the oral form of Copaxone.

Now she had returned to go over the consent form with her doctor, a necessary first step before joining the trial. She was a soft-spoken person, and seemingly calm about becoming a research subject. She had already read over and discussed the consent form with the research nurse, Sandy Cook. So the process was, from her point of view, pretty routine. The only question she asked had to do with whether she would be able to get Coral, as the oral Copaxone was called, after the trial ended, if the results were positive. Weiner explained that they couldn't promise, but added, "You would be the first in line for it since you've been in the trial."

Howard Weiner and his patient sat together at a small table in the examining room, going over the main points of the consent form. Sandy Cook stood, with arms folded, behind Howard, making sure he didn't, in his enthusiasm, overpromise. "This is the same medicine that's now being given by injection," Weiner explained. "But this is to avoid the injection."

"That's good," the patient said, with a quiet laugh.

He added, "We're excited about this because we worked on it right here in our laboratory."

"Really?" she said, animated for the first time. "Right here?"

"That's right," Weiner said. "This comes at the end of a fifteen-year effort. We had another trial that didn't work out. But we're hoping this one will."

About the Author

Susan Quinn is the author of two highly acclaimed biographies: *A Mind of Her Own: The Life of Karen Horney* and *Marie Curie: A Life.* She has also written a book about the theater, as well as articles for numerous periodicals including the *New York Times Magazine* and *Atlantic Monthly.*